Centre of Gravity

of

Gravity

THE PRINCIPLES SOLDIERS USE TO THINK, ACT AND ACHIEVE SUCCESS

JOSH FRANCIS

**Centre of Gravity: The principles soldiers use to think,
act and achieve success © 2019**
Josh Francis (Red Diamond Consultancy)

ISBN: 978-1-925962-83-3 (paperback)
The Camouflage Series – Book 2

Cataloguing-in-Publication information for this title is listed with the National Library
of Australia.

Published in Australia by **Josh Francis (Red Diamond Consultancy)** and
InHouse Publishing.
www.red-diamond.com.au

Printed in Australia by InHouse Print & Design.
www.inhousepublishing.com.au

Red Diamond
www.red-diamond.com.au
Follow us on **Instagram** and **Facebook**
Join the **readers group** for more publications

Please feel free to leave an honest review on Amazon or Goodreads

Special thanks to Cav –
a brother in arms and my key advisor for this book

Also by Josh Francis

Battle Rhythm – The military inspired
personal planning, discipline and motivation guide
(The Camouflage Series Book 1)

Pegasus – The Zach Kryton introductory short story
(Zach Kryton Book 1)

Coming Soon

Under the Pump – Anecdotes of a service station operator

Poseidon – The sequel to Pegasus (Zach Kryton Book 2)

Contents

Introduction

"We sleep peaceably in our beds at night only because rough men stand ready to do violence on our behalf."

— GEORGE ORWELL

Throughout history men and women of a certain breed have been renowned for their discipline, resilience, professionalism, and their ability to successfully achieve results in the world's most hostile environments, all while undertaking the biddings of their political masters. They are known as soldiers. The job of a soldier is unique amongst all professions in wider society, and combat soldiers in particular are trained in specific skill sets that enable them to act as part of a team, risking their lives to conduct the most dangerous tasks in the most difficult conditions.

The very nature of soldiering, which often involves demands and dangers rarely experienced in civilian society, requires thinking and action at a level much more considered and disciplined than virtually any other vocation. This is because a soldier, even at the most basic level, may be called upon to undertake activities that could result in the deliberate taking, or saving, of lives. The best armies in the world train frequently to prepare for these possibilities, simulating as closely as possible all the scenarios they anticipate having to face. The Australian and US armies in particular (and to no lesser degree their counterpart services) have been called upon, especially in the last two decades, to conduct a myriad of operations globally across the full spectrum of warfare, ranging from peacekeeping and humanitarian roles to intense and enduring combat operations. These two allied nations have a history of fighting together in global conflicts. From the jungles of New Guinea in World War II, to the frozen mountains of Korea, to the humidity of the Vietnamese swamps, and more recently to the dusty plains of Afghanistan, the Australian and US armies have many shared experiences. As a former Australian soldier, I have served extensively on operations and in training exercises with the US army. Hence, when I use the terms 'army' and 'soldier', I am talking about both of these nations' services as their experiences, which go back as far as World War I, have been very similar. Although they still have many cultural differences, these shared experiences have influenced their current training and doctrines, which are very similar and which allow for high levels of interoperability during training and operations. As a result, they have produced the soldiers whose skills and knowledge are so eagerly sought.

To ensure that it can conduct operations consistently – and successfully – the army trains its soldiers to the highest possible level in the skill sets they need to execute a range of operations, using well-defined principles which are the foundation for training and are utilised in combat. Soldiers are trained not only in physical skills but also in the mental and academic skills that are required to conduct operations in complex and ambiguous environments and that instil the attitudes needed to successfully operate in such environments. The principles ensure that soldiers become well rounded, highly trained individuals who can work as part of a coherent team.

This combination of training and the experiences gained from over a century of warfare is why the army is widely viewed as a highly professional and resilient force, respected by allies and adversaries alike. A professional force is a direct reflection of the calibre of the individuals who make up its numbers, and soldiers are renowned not only for the service they provide but also for the many skills that they gain during their service.

Soldiers always stand out from the crowd. The way they dress is a reflection of the discipline instilled in them. The way they walk and comport themselves, usually with a sense of purpose and authority, exudes confidence and is a reflection of the responsibility placed on them as members of the profession of arms. The very best soldiers are friendly and engaging, yet methodical, logical and considered in everything they do.

There has been a recent trend towards former military personnel who have left their service finding success consulting to businesses in areas of organisational management and leadership. Australian and US soldiers who have significant operational experience, particularly through operational service in Iraq and Afghanistan, have been sought after by companies who recognise the 'soft' skills that former soldiers bring to the workplace and the underlying principles that helped build these skills. Soft skills, such as a strong work ethic and attitude, loyalty, high levels of discipline and motivation, and the ability to think critically, be resilient in difficult situations and work comfortably in demanding environments, have in many cases made army veterans preferred employees by both public and private organisations.

Their abilities are valued because employers are looking to leverage the strong reputation the army first established during the 20th century and enhanced recently as a result of enduring operations in the Middle East and around the world. The success the army has achieved is a direct reflection of the personnel who serve within it. The vast majority of these soldiers are young men and women in their 20s and 30s. No other profession places so much responsibility on people so young. A standard combat infantry platoon[1] comprises a young lieutenant who is likely to be 22 or 23 years of age. The soldiers under the lieutenant's command have an average age of 19 or 20 while the platoon sergeant, the second-in-command, may be in his or her early 30s.

Soldiers are a reflection of the society they serve and protect. They share the common bond of having the desire to serve their nation. Much of the knowledge, skills and attitudes (I will use the broad term 'skills' to describe these three aspects) that soldiers are instilled with and further enhance throughout their careers can be applied to your own life and to

---

[1] A platoon is made up of 30 soldiers, divided into 3 sections of 10 soldiers.

your personal endeavours, resulting in the ability to approach the things you do in a more efficient, professional and effective manner. They are principles that can help establish a foundation for how you want to approach things in your own life. Just like soldiers, who come from all walks of life, learn how to think, act and succeed in all the endeavours they undertake in uniform, you also have the ability to improve yourself by thinking and acting like a soldier. By harnessing the skills soldiers learn, you can develop a new approach to everything you do that will allow you to become a more confident person. It doesn't matter whether you're a white-collar professional, an athlete, a student or a stay-at-home parent.

Success can be both subjective and objective. Objective success is about whether common, fixed goals have been achieved or not, such as making one million dollars in revenue for a company or being able to do 50 push-ups. Subjective success for one person may not be considered success in the mind of another. A personal best time in a five-kilometre run for one person may not be the personal best time for another, so they will view achieving success differently. It's a case of what goals you're setting for yourself and whether you're achieving them. You have to be critically honest with yourself but also fair about what success looks like in your endeavours.

This handbook, the second in *The Camouflage Series*, serves to act as a personal development guide. It will encourage you to instil in yourself some of the characteristics of soldiers, to think and act in a way that will allow you to improve your personal and professional self and to achieve success in all your pursuits. It will not turn you into a stereotype soldier. You won't suddenly be waking up your household to bugle calls or leopard crawling to your car to avoid snipers! Quite the opposite. It's about helping you develop the skills that make the army, and its soldiers, stand out from the crowd. Regimentation has a time and a place in the army, but unlike some movies would have you believe, soldiers don't march around everywhere in cadence, say 'Yes, sir,' at the end of every sentence or sleep at attention in ironed pyjamas. Using various examples, I will discuss and describe how the army ensures that its soldiers have the key skills needed to thrive in complex and dangerous environments and how you can apply them to your own life so that you can better achieve the things you want to achieve by approaching them in the same manner that soldiers do. It will help develop your professionalism and improve your personal outlook on life so that you can approach work, fitness, relationships or any undertaking with greater confidence and achieve more success and fulfilment.

It will help you think, act and succeed like a soldier!

The Centre of Gravity

"A certain centre of gravity develops, the hub of all power and movement, on which everything depends."

— CARL VON CLAUSEWITZ (*ON WAR*)

As stated in the introduction, soldiers think and act differently. They're taught to. They learn to. They need to. They need to think and act in a way that most people never normally will because soldiering can be unforgiving. It is not compassionate to avoidable mistakes, and it doesn't tolerate mediocrity. Lives often depend on how well soldiers carry out their duties. As a result, they have to give consideration to risks, consequences and other influencing factors that aren't typically found in the civilian environment. The very nature of their job requires an approach, in their thinking and their actions, that sets them apart from their fellow citizens. Their training, the experiences they have gained during their service and the level of responsibility and trust with which they are comfortable and experienced working under are all factors that have led soldiers to obtain the unique skills that are highly regarded and sought after.

Soldiers achieve success through a combination of factors – good training, good equipment, the use of supporting elements (such as medical and logistical support) and sometimes sheer good luck. However, perhaps the most critical reason is because they have a firm belief in what they are doing and in the principles that their country, their army, and they themselves stand for. I will not attempt to delve into all the political or cultural views and ramifications pertaining to the involvement in Afghanistan and Iraq since the September 11, 2001 attacks on US soil. Despite many people's personal beliefs on the rights and wrongs of these commitments, almost universally agreed upon is the fact that the soldiers who were deployed to these war zones performed their duties in an outstanding manner and often achieved great things against almost impossible odds. This has widely been put down to the fact that the army is an all-volunteer force and that the men and women who wear the uniform make a conscious decision to undergo the rigorous training and subsequent deployments that could potentially get them killed.

Each soldier joins and serves in the army for different reasons. These reasons can often change over time. Soldiers are only human, after all, and suffer and experience the same emotions as everyone else. The reasons for joining belong to the individual. I have heard comments suggesting that some people join the army because they can't do anything else in life. After my own 18 years of service, I truly believe this sentiment to be completely inaccurate and based on historical fallacies. Modern soldiers are technologically savvy and many have tertiary qualifications.[2] Many have completed multiple deployments. The level of responsibility, often grave responsibility, placed upon young soldiers in modern warfare is proportionally higher than that of their predecessors in previous conflicts, partly due to the need for soldiers to be able to master multiple skill sets in order to conduct operations in rapidly changing operational circumstances.

This model was the result of the development of the 'three-block war' concept, a US notion developed in the early 1990s. This concept put forward the theory that the training of junior soldiers for modern warfare required

---

[2] In 2012, former US soldier Nick Palmisciano authored an essay where he outlined that only 0.45% of the American population had served in the Global War on Terror, without the war having any noticeable difficulty on the civilian population back home. Additionally, his research suggested it was harder to enlist in the military than it was to gain access to college, based on higher levels of physical and academic requirements.

them to learn and have the ability to operate along the full spectrum of military operations, including full combat, peacekeeping and humanitarian actions, often within the same geographical area and potentially within days and hours of each other. The concept was designed and developed to ensure low-level unit leaders would be able to make major decisions and take action that might have serious strategic ramifications. This concept led to the coining of the term 'the strategic corporal'. In Afghanistan, for example, it was not uncommon for a corporal or sergeant to lead a platoon of soldiers in combat one day and then conduct a meeting with village elders to determine community infrastructure needs the next, knowing full well that some of the locals at that meeting had probably been involved in the previous day's firefight. This combination of skills, training and equipment and the physical, psychological and moral strength which soldiers need to succeed is known as the centre of gravity.

Virtually everything has a centre of gravity. In technical terms, a centre of gravity is the strongest point or area that allows something to work properly. Removing the centre of gravity will cause it to be weakened or fail altogether. For a vehicle, for example, its being on all four wheels (centrally aligned). Having the engine running and propelling the vehicle along is its centre of gravity. When that vehicle speeds up and makes a sharp and sudden turn, there is the potential that it will overturn because the momentum and inertia force the centre of gravity to be weakened. All the things that make the vehicle strong and capable (i.e. the wheels on the road and the engine turning them) are no longer in place, and the vehicle may end up flipping over. In a non-technical situation, a centre of gravity is still about the facets that allow something to remain strong and work properly. For a sporting team, the centre of gravity might be a few key individuals who are playing and leading the team every game because they have a mix of talent and leadership skills. If those players are not on the team, whether due to injury or because suffering poor form, then the centre of gravity for that team will be weakened, and another less talented team can defeat them. This is why sports teams always prefer to have a *champion team* instead of a *team of champions*, because the centre of gravity of the former (in this case, a larger group of players who work and play really well together so that the team doesn't rely on just a few individuals) is much stronger and can withstand a few injuries to their players because the rest are able to still play effectively and cohesively and haven't become reliant on the key few players.

In both Australian and US military doctrine, a centre of gravity is generally defined as *the source of power that provides moral or physical strength, freedom of action, or will to act.* Essentially, it's about what makes an army strong and gives it the ability to be successful on the battlefield. These facets include things like equipment, training and personnel (inclusive of sub-factors such as skills and morale). Military planners will attempt to determine what the centre of gravity is for formations of troops and equipment, both their own and their enemies' (which, respectively, need to be protected or destroyed), as this allows for a better understanding of respective strengths and weaknesses. A centre of gravity may have many facets, not necessarily just one. A centre of gravity for a military force may be its technologically advanced equipment as well the level of training of its soldiers. Without these facets, a military force is not as strong as it could be, and its ability to conduct operations might be limited or particularly vulnerable to enemy action. Planners will then need to determine what resources the force needs to have or acquire in order to conduct the mission effectively, as well as determine where the enemy vulnerabilities lie so that they can be exploited.

A centre of gravity is not always equipment or personnel. During the war in Afghanistan post the September 11 attacks, both Australian and US forces (and other allied nations that formed the coalition) quickly identified that the centre of gravity for both their own forces and the Anti-Coalition Militias (ACMs)[3] was the civilian population. Because of the nature of asymmetric warfare, whereby there is no discernible front line, having the support of the civilian population was critical in ensuring that there

---

[3] The term ACM was given to the loose alliance of all anti-coalition elements, which included the Taliban, Al-Qaida and other terrorist forces.

was freedom of movement for coalition forces because it meant that these towns and villages wouldn't allow the ACM the ability to hide among the general population. This was achieved, with some exceptions, by going out into the towns and villages and helping the communities, and in some cases, living in the nearby area through the creation of forward operating bases (also known as combat outposts) which were scattered across the countryside. This help could include building schools and hospitals, providing education and medical training, or training the local police forces to provide security in the absence of coalition soldiers. The ACM took the same approach but instead utilised fear and violence to intimidate the population into supporting them or coercing the undecided villagers into choosing a side, often committing horrendous atrocities against those people who supported the coalition.

A similar approach was taken by US forces in Iraq where commanders engaged with local sheikhs and tribal leaders to seek their support to fight Al-Qaida and the myriad of vicious Islamic extremist groups that were the early iterations of ISIS (also sometimes referred to as Daesh). The US commanders conducted an assessment of what their centre of gravity would be and quickly realised that they had to seek the support of the local communities because they knew that the massive firepower of the coalition armies alone would not be enough to successfully conduct warfare in an asymmetric environment. They assessed that success would be achieved by getting the civilian population onside and, in turn, being able to operate more freely with local support.

Determining your own centre of gravity allows you to better understand where you can draw your strength – physical, mental, moral and professional – and in turn gain the confidence, self-belief and sense of purpose to do the things you want to do successfully. It's those tangible and psychological things that empower you, give you self-esteem and help you to achieve your goals. This is important for your own personal and professional development because a strong centre of gravity allows you to rise to the challenges you face in all your endeavours, just as soldiers do. It is about knowing and using the things in your life that make you feel strongest. This will equip you with the confidence, self-belief and sense of purpose needed when you're required to step out of your comfort zone so that you can do the things you need to do, or that you thought you weren't able to do.

Most soldiers would argue that their own personal centre of gravity is their belief that what they are doing in operations is morally right, including the strong bonds formed from serving with their fellow soldiers who share the same motivations, and their ability to work with cutting edge technology and being highly trained in the use of it. These are moral, physical and psychological capabilities and strengths. These allow the soldiers to be able to do what they need to do to conduct their missions. The training and experiences soldiers receive almost universally results in their developing three key internal attributes which set them up for success in their army careers as well as later in life: confidence, self-belief and a sense of purpose, i.e. the confidence to deal with adverse situations and to deal with difficult people; the self-belief to venture out and achieve the goals they set for themselves, to take knocks and get back on their feet and keep going; and a sense of purpose in that they are doing something worthwhile by contributing to their country in particular and to humanity in general. Confidence, self-belief and a sense of purpose are the result of a soldier's strong centre of gravity. This itself becomes an additional source of strength and is why they can deal with operating in the most atrocious conditions, putting their lives on the line and making huge sacrifices that impact them and their families. It's how a modern army can have success on a foreign battlefield many miles from home.

Soldiers are well known for their confidence. It permeates their being, revealed through how they dress, how they walk and how they talk. It comes from having undertaken extensive training and endured suffering with tests of their physical and mental strength that most civilians will never appreciate or have to endure. All soldiers go through basic training. In US parlance, it's known as 'boot camp' whereas in the Australian context, it's referred to as 'recruit school'. One key training method the Australian army uses to instil confidence in recruits is called the high-wire course. This is a series of obstacles, strung up several metres above the ground, that recruits have to navigate by walking along a metal wire whilst tethered to a safety harness. Many of the obstacles require the recruits to jump across platforms affixed to large wooden poles or to hang precariously above the ground while they navigate ropes or monkey bars using only their arms. Despite being safely tethered to their harness held by ropes which are managed by highly trained staff, it is often a frighting experience, especially for those fearful of heights. The experience forces many of the recruits to step out of their comfort zones for the first time in their lives. Once completed, the recruits have a better understanding of their own capabilities and can draw upon those experiences when faced with other challenges. Their confidence in themselves increases significantly. It improves their self-belief as they realise that they have successfully achieved something that they probably thought they couldn't do. Their sense of purpose is reinforced because, in order to graduate and move into the wider army, they have to complete the course.

Soldiers will then, depending on their chosen career paths, undertake additional highly dangerous and demanding training. This training will further reinforce their own confidence in themselves and enhance their self-belief and sense of purpose in what they are doing. This ensures that, if or when they are in combat or in seriously adverse and dangerous situations, they have the belief they can deal with the situation, assess the problem to make a decision and act in the appropriate manner to achieve the outcome required. They can utilise the skills they have been taught and look inside themselves to draw strength with a calm and disciplined approach when dealing with the 'fight or flight'[4] type of situation.

---

[4] This is a concept first coined by Walter Bradford Cannon that the body will have one of two physiological responses to stressful situations: Fight – whereby a person will attempt to defeat or face the perceived threat; or Flight – whereby a person will seek to remove him- or herself from the stressful situation.

Just as army commanders look to make an assessment of their own forces so they know where to best use them (and protect them) in order to undertake missions, knowing your own centre of gravity will allow you to approach life with a better appreciation of what you do well so you can leverage your strengths to be better able to work towards your goals. By thinking about this concept, you can determine where your own strengths lie, what contributes to and supports your self-belief and sense of purpose and, ultimately, from where you derive confidence. You can then consider how to best leverage that confidence, self-belief and sense of purpose to do the things you want to do and achieve the things you want to achieve, whether it be at work, in the gym, when dealing with people in the course of business or in your relationships.

Everyone has something, perhaps several things, which makes up his or her own centre of gravity, and from which he or she draws confidence, self-belief and sense of purpose. Some of these are material, others physical, personal or spiritual in nature. A significant traumatic experience that someone has gone through, and recovered from, is an example of something someone could draw strength and confidence from. Such people know they have the ability to deal with difficult circumstances, and when facing challenges in future endeavours, they remember the time in the past when they overcame adversity and use that as the impetus to face the next challenge, confident in their own abilities. Suffering a significant setback in an endeavour can motivate people to try again and strive for success. An Olympic athlete who sees a rival win the gold can be given a renewed sense of purpose to train even harder for the next Olympics.

Being wealthy gives some people the freedom to pursue things without financial limitations, as it allows them the freedom to buy the best equipment or to get the best training. They become confident in the knowledge that they won't risk wasting money on pursuits that might not work because they have the means to try multiple times. This is a material strength but, unfortunately, not one available to the vast majority of us. However, being diligent with personal finances and having learnt good cash management skills from friends and family, or even self-help books, can give some people the self-belief that they can manage a large home loan, even if on a moderate wage.

A strong family or friendship base is another place people draw strength from. Just like soldiers, who rely on their brothers-in-arms to protect and support them, people draw confidence from knowing that they have friends and family to give them tangible and moral support in times of crisis or difficulty. Their family and friends may also encourage them to do things that they might not otherwise do, such as to apply for a job or to begin a weight loss program. Many people draw great strength from their husband, wife or partner because they gain confidence from having someone to lean on and discuss matters with. I have served with many soldiers who always said that the reason they spent so much time away from their wives and children was because they wanted to help make the world a safe place for their children, and they were prepared to sacrifice time away from their families to help contribute to this. When things became difficult during operations and they started questioning whether the effort and time was worth it, they drew strength to continue the mission from knowing that they were doing it for their children (and the children of many other people, too).

This is why it's wise to often question who your real friends are. Do you have friends who will literally drop things at a moment's notice to come and help you if needed? Or do you have a lengthy list of Facebook-type friends who are more acquaintances rather than true friends? Life keeps us all busy. It's part of living in the modern world, and even though I might not see or hear from some of my friends from my time in the army for months at a time, I have complete confidence that they would be there for me should I need them just as I would be there for them. It's a loyalty based on having spent multiple deployments around the world together and having shared the same difficult experiences. I draw confidence in knowing that, in anything I do, I have a small group of loyal friends who will help me should I get into difficulty. Unfortunately, you sometimes only find out who your true friends are when you really need some support. It's the ones who are there for you who are your true friends. These are the ones from whom you draw strength and support.

Fitness is something which all people should incorporate into their lives, as being fit has both mental and physical health benefits. Soldiers and scientists alike know that being fit gives you the energy boost to deal with the general rigours of day-to-day life, and a good workout will release endorphins that promote clear thinking and physical strength and build confidence. Fitness also reduces the chance of injury or illness. Being fit is an inherent requirement for all soldiers, particularly combat soldiers, as it is what allows them to fight wars in environments ranging from tropical jungles to scorching hot deserts and to keep fighting even when they begin to fatigue. Feeling fit boosts confidence as being fit requires discipline and commitment, and these traits can then be applied to anything you do because you will feel physically and mentally strong as opposed to lethargic and full of self-doubt.

Have a think about your own centre of gravity and what you can to do to make it stronger. Think about the things that give you strength and that in turn give you confidence, self-belief and a sense of purpose. When do you feel strongest and most confident in your abilities? Is it after a good gym session? Is it after sharing your concerns and troubles with friends whom you respect and who always give you sage advice? Or is it after seeing an inspirational movie or reading a story? Perhaps it's from thinking about experiences in your own life that you successfully dealt with despite adversity that made you a better person. What are the things you do well, or have done well in the past, that give you self-belief and that you can leverage to approach things differently or in a more efficient way? What are your goals or ambitions that motivate you and give you a sense of purpose to achieve things or live a more fulfilled life?

They don't even have to be related to the things you're working towards. For example, you might set some fairly ambitious financial goals and draw confidence and self-belief that you're disciplined enough to achieve them because you spent your childhood getting up early to undertake swimming practice and have learnt that patience and discipline pay off in the long run. You might have to deliver a work presentation to a large audience and draw strength to overcome your shyness by thinking about how you're an excellent runner who has been able to run marathons and that no one else in the audience has the ability to do that. You're thinking about other things, not related to public speaking, that have given you confidence and self-belief. Apply that type of thinking to everything you do, as knowing you have a strong centre of gravity leads to enduring confidence, self-belief and a sense of purpose. Also, look at the tangible things that give you a strong centre of gravity but that you might have overlooked or just taken for granted. Having a home, being well fed and in good health might be part of your centre of gravity because the fact you have these things means you have the ability to set out to pursue goals and don't have to focus on finding shelter or wondering where your next meal will come from or having to deal with ongoing illness. You have to look at all the things that exist in your life that allow you to do the things you want to do – physically, mentally, morally and professionally. This is your centre of gravity.

Where you identify areas of weakness or shortcomings in your life that undermine your ability to have a strong centre of gravity, you should set out to work on them. Soldiers attempt to become experts in the skills they are least confident in, so that when in combat they are not second guessing themselves by doubting their own abilities. They spend hours on the shooting range practising marksmanship or walk many kilometres practising land navigation and conditioning their bodies to ensure they are as well rounded in their capabilities as they can be. If you're not currently fit, start a physical training (PT) program. Even if time is tight in your routine, you can always seek to devote some of your week to doing fitness training. Anything is better than nothing. If you feel you lack confidence generally, try something that forces you to step outside of your comfort zone. Perhaps skydiving, a dancing class or trying a new sport can help you build your confidence base and your own belief in your ability to achieve things. Do it with a friend, leveraging your support network to build confidence. People who struggle with public speaking can join clubs like 'Toastmasters' which teach and encourage people to give small talks aimed at improving their confidence when speaking to other people.

If you aren't as comfortable in your relationships as you would like to be, make a concerted effort to work on them and resolve any issues or concerns that undermine the strength of that relationship so that the relationship is a strengthening feature of your life rather than a vulnerability. If there are areas at work that you want to be better at, study them and practise them so that you become more proficient and capable, armed with the skills that will make you eligible for promotion and allow you to become a vital part of your organisation's structure. If finances are tight and it seems like you never have any spare money, look to make a budget and review whether you're wasting money on things you don't really need. Look at ways you can better manage your money so that you can feel more secure in your financial situation and still enjoy doing the things you want to do. Additionally, seek to reduce costs by looking for cheaper options for utilities or a better rate for a home loan.

Unlike in battle, no one will be actively trying to defeat your centre of gravity, so the only person who can undermine it is you, which will happen only if you allow yourself to stop believing in your own abilities or if you lose confidence in yourself. That's why understanding your own strengths and weaknesses is vital to developing yourself personally and professionally because you can leverage your strengths to help you be successful in all of your endeavours and work on mitigating your weaknesses where there is a risk they will hinder your success.

Understanding and defining your own centre of gravity is a small but vital part of better understanding yourself and assists you in becoming a strong-willed person who can seek to face and overcome the challenges that are prohibiting you from achieving the success you're after in whatever area of your life.

Seeking to determine and strengthen your centre of gravity allows you to begin walking the path of personal development towards achieving success in your personal and professional life. I will talk in the next chapters about the other characteristics of soldiers that enable their success. An army's measure of success is essentially achieving everything that it sets out to do. Sometimes these goals are very ambitious and need to be amended as circumstances change. However, the army will always review its centre of gravity to ensure that it has the strength to undertake the missions asked of it and is best placed to be successful in all operations, whether they be assisting the local community during natural disasters or conducting complex warfare on the other side of the globe.

SUMMARY:
- **A centre of gravity is defined as** *the source of power that provides moral or physical strength, freedom of action, or will to act.*
- **It is about understanding your personal strengths and how these give you the confidence, self-belief and sense of purpose to deal with the challenges in your life.**

Professionalism

"We're not a profession simply because we say we're a profession."

— General Martin E. Dempsey

Everything the army does is a reflection of the professionalism of the soldiers within its ranks. It's apparent in the way soldiers dress and in the way they comport themselves when conducting duties. Professionalism is generally defined as 'the competence or skill expected of a professional.'[5] This is a simple definition, and one that is very true, but what does it *actually* mean to be a professional? In military circles, it's a combination of the knowledge, skills, behaviours, attitudes and abilities that are unique to soldiers. To be professional as a soldier means being highly proficient at your job and having the physical and academic skills expected of your rank and role. It means comporting yourself in a manner that brings a positive reflection on you and your fellow soldiers and having the maturity to control your emotions despite having to deal with complex situations and people who oppose and actively try to undermine everything you stand for or are trying to achieve (i.e. the enemy). It means undertaking every task with a methodical and considered approach to ensure the desired outcome and to instil faith and trust in fellow soldiers as well as ordinary citizens. It's about approaching soldiering in a business-like manner, at a level beyond what other organisations might consider is required to complete the job. Above all, it's about not resting on laurels. The army will consistently strive to maintain and promote its professionalism, achieved through how

---

[5] Oxford Dictionary, www.oxforddictionaries.com

it conducts business and how its individual soldiers perform and behave. General Martin Dempsey, the 18th Chairman of the US Joint Chiefs of Staff, on speaking about the US military, once stated that "we must continue to learn, to understand and to promote the knowledge, skills, attributes and behaviours that define us as a profession." He was stating that professionalism needs to be about enduring practices.

Professionalism enables success. It's about the manner in which you approach all the things you do in your life. It's about doing everything you do diligently, calmly and with the intent to do it proficiently and effectively, which will allow you to achieve things you want to achieve. In short, it's not just about striving to be good but to be outstanding, both in your personal and professional life. You don't have to be an actual professional (i.e. be paid for doing something and be one of the best at it) in something to have a professional approach to it. Whether it be at work, in the gym or in the way you renovate your backyard, a professional approach will ensure that you're giving your best effort in all that you do, which will result in your achieving the things you need and want to do.

STANDARDS

There are numerous ways that the army instils professionalism in its soldiers. The key one is the setting of standards. Standards are the required level of effort and proficiency expected of soldiers in all the tasks they undertake. These can range from how personal weapons are handled and fired in combat to how uniforms are ironed and worn to how a tent is raised and operated while in the field. Combat soldiers will position all their equipment in certain pouches on their webbing rigs (the vests that hold body armour and where they place their weapon magazines, grenades, knives, etc.) so that, if one of them is injured, his or her fellow soldiers instantly know where to obtain the first aid kit or to find extra bullets if needed. Standards are set to ensure everyone knows the expectations required of them as soldiers, which allows for uniformity. This uniformity builds the foundation of a professional force.

Standards mean not doing just the bare minimum that is required (to use another term, to not do things 'half-arsed') because that risks losing the trust not only of other soldiers who do have high standards but also of the taxpaying population who expect professionalism from their army and its soldiers. Setting standards outlines a level of effort and proficiency which becomes the baseline for all that soldiers do. This ensures that if they are ever in combat, they will be used to the demands and rigours which they have learnt and practised during training and won't suddenly be trying to perform at a higher level that they are not conditioned for. It means that they have the highest level of proficiency in the skills that will protect them and their fellow soldiers and lead to success in the battle.

If soldiers can't initially meet the standards that are set by the army, especially during recruit training, then they are given every opportunity to reach them, whether they be the standards in the soldiers' chosen trade skills or in common areas like physical fitness. Many civilians join the army and enter recruit training well below the fitness level they require to graduate, and many have never held or fired a weapon in their life. The army knows full well that it recruits soldiers who need to be rigorously trained in order to meet the basic standards needed to be an effective team member of the various units of the army, and so it trains them accordingly. As a soldier's career progresses, he or she will undertake further training in skills that require very high standards, and many will not be able to meet these requirements. That doesn't make those soldiers any less professional; it's simply that higher standards will result in higher levels of professionalism because the very nature of service in special forces, as with many other parts of the army, is highly demanding and is unforgiving of mediocrity. Entering special forces is a highly sought-after career path for many soldiers. However, it requires the highest standards of physical fitness and mental aptitude, and many soldiers do not make it to the ranks of those elite warriors.

The Australian army has an annual fitness test called the *basic fitness assessment* (BFA). Each soldier is required to do a series of push-ups and sit-ups and then a 2.4-kilometre run or a 5-kilometre walk. It is not an overly difficult test, and most soldiers can pass it if they're reasonably fit and if they're not injured or unwell. Each soldier is required to meet different standards based on their age and gender (the cause of much discussion and opinion amongst all soldiers!). For each age and gender group, there is a range of incentive levels which sets a higher standard for individuals to personally aim for. As an infantryman, especially one in an airborne unit, I was always expected to be able to complete the highest incentive levels, which were to do at least 60 push-ups in two minutes, 100 sit-ups to a three-second cadence, and to run the 2.4 kilometres as fast as possible. (Any time under ten minutes was considered a good effort.) This was the airborne infantry's desired standard because it was the entry requirement to undertake parachute training. Even after I moved on from the regular army to the reserves, I always attempted to achieve these levels - not only as a matter of my personal professionalism but also because it was good leadership. How could I ask anything of my soldiers that I wasn't prepared to do myself? I expected them to meet high standards, and I had to be able to demonstrate them. Often it wasn't about doing the most push-ups but simply meeting the expectation of giving your absolute best. In the latter part of my career, I took great umbrage at older officers and senior non-commissioned officers (SNCO) who would do the bare minimum in the BFA and then pompously brag in front of younger soldiers about how they didn't have to achieve the higher standards anymore. This was very unprofessional, and I would often tell them later (subordinately) in private what I thought of their attitudes and actions.

Many sporting teams take a professional approach to how they conduct business, even if they are not in a professional competition. I am amazed by the level of professionalism with which college football teams in the US approach their sport. They have highly tuned recruiting capabilities, wear uniforms supplied by big name brands, and their stadiums rival those of professional sports teams in terms of size and facilities. Yet all the players are unpaid, full-time college students (scholarships notwithstanding). However, it's not just looking professional that makes these teams highly regarded, but it's predominantly their actions. Players on college teams commence an off-season training program many months out from the playing season, and they have practice schedules that rival those of NFL teams in intensity. This training is complemented by a full-time study schedule and often by commitments they're expected to undertake even when not on the gridiron. It's no coincidence that the most successful teams are the ones that take the most professional approach to their sport, even if they don't have the best facilities or the most talented players. It's because they set high standards for themselves. There is an expectation for players to meet those standards in everything they do, on and off the playing field. Anyone falling short, whether because of performance or behaviour, is dropped from the team.

Martial arts students have standards instilled in them when they start training, and the standards are maintained long after they achieve a black belt and start instructing students themselves. The standards apply not only to the techniques and skills they need to achieve in order to progress to a higher level but also to how they dress, behave when on the training mat and treat their superiors and subordinates. Many martial arts students are taught to take the lessons they learn, specifically related to having high standards, into their personal lives.

Having high standards is about being methodical and meticulous with the things you do, without being obsessive-compulsive. It ensures that there is integrity in what is being done. The standards the army sets are high because they ensure that soldiers are highly trained and can deal with all the challenges they may potentially face and are therefore confident in their own abilities, as well as in those of the soldiers around them. It's the little things that may seem abnormal to ordinary people that simply become part of the normal routine for soldiers. Weapons are cleaned and oiled every day, even if they haven't been fired. This ensures the soldiers know that they are in the best possible state, ready to be employed if needed. In football terminology it's called doing 'the one-percenters'. These are the little actions and efforts performed during a game that make the difference between a good player and a great player.

Leaders are responsible for setting standards and ensuring that they are met. This guarantees the efficiency of the unit and maintains professionalism. The Australian army has a saying: 'The standard you walk past is the standard you accept'. This was coined by a former Chief of Army[6] as a means of saying that if you're not prepared to correct the poor standards, then you obviously accept them and will be held responsible for them.

Professionalism is the by-product of setting and maintaining high standards and being disciplined enough to consistently meet them. Although professionalism is instilled in all soldiers, it still ultimately comes down to the individuals and how ready and willing they are to conduct themselves in a professional manner. The best units in any army, those that are renowned for their professionalism, are made up of soldiers who have solid leadership and who personally set high standards for themselves, even above the basic requirements.

Setting high standards will ensure a professional approach in everything you do because it means the manner in which you do things will not only look professional but actually be professional. We've all experienced bad customer service when shopping, where a call-desk operator or a shop assistant has been rude, disinterested or incompetent. It's usually enough to make us reconsider engaging those services again. Now think about the best experiences you've had, the ones that made you feel like you were dealing with a professional. These experiences were positive because the people serving you made the effort to do the little things to make your experience more fulfilling and more enjoyable. These are the people who did the 'one-percenters'. Not only did they look and act like professionals, but they knew their job well. They obviously had high standards set by their workplace or had personally chosen to have high expectations of themselves (which is why experiences may differ when shopping at the same place).

---

[6] 'Chief of Army' is the professional head and the highest ranking officer in the Australian army – a three-star Rank (Lieutenant General).

As individuals, we are responsible for maintaining the standards we set for ourselves and critically analysing whether we are meeting them or not. The standards you set for yourself need to be realistic, however, as they are different from goals. For example, you might want to be able to bench press 100 kilograms. This is not achievable for many people, so in the absence of the ability to do this, your standard might simply be to maintain a regular gym routine and to ensure you do all your sets and all your reps every session with the correct technique and not take shortcuts during your session.

Look at the standards you have set for yourself in the things you do, whether they be in your workplace, your gym routines or even maintaining your house. Look at how you can do the 'one-percenters'. What can you do better to become more professional? What small things can you do that will allow you to do things more efficiently and ensure you can better achieve your goals? Perhaps it's just the little things in life that make you a better citizen, a better partner to someone or just make you feel better about yourself. It's as simple as making a phone call to confirm a meeting with a client even though it's not required or expected or putting your weights back when you've finished a gym session so that the next person is able to use them. It's as simple as considering the people around you when doing things that might impact others, such as turning your phone off in the cinema or not having a loud and obnoxious conversation on a bus. In the workplace, be on time to meetings, or make it clear to colleagues if there are things that will affect them so that they can better prepare their own day. The occasional purchase of flowers for a loved one or a phone call to check in on a friend are all small things but can make a difference in someone else's day. Doing the 'one-percenters' for the small things will ensure that you're well versed in doing them when it comes to the big things, like potentially making billion-dollar deals in the corporate world. These simple things will go a long way in contributing to your own personal and professional development but also in how people view you and whether they want to interact with you. The standards you set for yourself are a reflection of your own professionalism.

VALUES

Values are principles. They are one's judgement of what is important in life.[7] Values are the *personal* standards which are instilled in soldiers – the individual expectations of each soldier, which collectively combine to create an army which has shared values. They are the very foundation of and guide for how the army comports itself and how it wishes to be characterised. Having values, and using those values as a guide towards the way it operates, is perhaps the key to the army's professionalism. They are the principles which guide soldiers' conduct in battle and in peacetime and ensure that, in the ambiguity and fog of warfare, soldiers have guidance to help them make the right decisions and not succumb to the same behaviours and actions that many of their adversaries do.

The Australian army has the following four key values:
- Courage
- Initiative
- Respect
- Teamwork

The US army uses the acronym LDRSHIP to illustrate the values for which it stands:

L – Loyalty
D – Duty
R – Respect
S – Selfless Service
H – Honour
I – Integrity
P – Personal Courage

While these are the listed core values of each army, both would agree that all these values apply to their soldiers - a reflection of the values

---

[7] Oxford Dictionary, www.oxforddictionaries.com

both their democratic societies were founded on. Any infringement that undermines or contradicts these values is taken very seriously, regardless of whether it occurs in war or peace. The US army took serious action against the soldiers who perpetrated the prisoner abuse at the Abu Ghraib prison in Iraq in the aftermath of the 2003 invasion. Similarly, the Australian army took action against several soldiers from a Townsville-based infantry battalion who took photos of themselves dressed in white hoods that resembled Ku Klux Klan garb in the early 2000s. The fact that these soldiers were heavily involved in overseas operations at the time was no excuse for the actions of the few individuals involved in these rare cases.

Both these incidents went against the values that the army had set for its soldiers, and the perpetrators were punished accordingly.[8] Values are important for the army because they ensure that professionalism can be maintained among the ranks, and they help promote a common purpose based on a belief system. They give soldiers guidance as to what expectations are required of them as individuals and allow unit commanders to enforce standards among their units which are common across the entire army. This allows for cultural uniformity as all soldiers have signed up to adhere to the same values and beliefs.

The values are what ensure that soldiers treat newly detained prisoners of war with respect, even after a firefight that may have killed their mates. They're what ensure that a soldier can confidently differentiate between an order from a superior that is unlawful and should not be obeyed and one that is within the laws of armed conflict and the established rules of engagement. The values guide soldiers in how they comport themselves at all times because they understand that they are representing their nation. Most of all, values are what give soldiers the firm belief that they have the moral right, so that when they are called upon by their government to

---

[8] The majority of the soldiers received internal disciplinary action, whilst a small few had their enlistment terminated.

potentially take lives, they are validated in what they are doing and that it is in line with the expectations and beliefs of the wider society.

Personal values allow you to articulate the type of person you want to be, personally and professionally, and to better understand what is important to you in your life. Having values promotes professionalism because they act as the guide for how you live your life, adhering to a set of principles that you set for yourself. Your behaviours and attitudes, in line with your values, will be noticed by those around you and will instil confidence in and respect for you in others.

Have a think about what your key values are. You probably have lots. Values can be described in many ways, using phrases or single words to define what you believe in and what you let guide you in how you want to live your life and treat others. You'll find very quickly that the people who are your closest friends, those whom you get on well with and those whom you most respect and admire, share these same values. Values are the framework for how you live your life and how you would like people to notice you and characterise you. They help you achieve success because in everything you do, in every decision you make, you ask yourself whether it adheres to your values and whether it allows you to achieve the things you want to achieve and become the person you want to be. Having set values helps you better understand your own centre of gravity, as they are the foundation of who you are and what you want to be. Take pride in them, because when life becomes difficult and you start to think you're losing your way (as soldiers sometimes do after a lengthy deployment and when away from home frequently), you can refer to them as guides to remind you what is important in your life and why you're doing the things you are.

DISCIPLINE

In the first handbook in The Camouflage Series, *Battle Rhythm*, I spoke in great detail about discipline. All soldiers are instilled with discipline from the moment they step off the bus at recruit training after enlisting.

Discipline is the ability to push through difficult circumstances, to avoid distractions and to not take shortcuts even when they might be available and no one is looking. Discipline is what allows you to commit to a course of action and persevere even when you begin to face obstacles or conditions become unfavourable. You need discipline to commit to a weight loss program and keep doing the physical activity required to achieve your weight loss goals. You require discipline to finish an essay that is due on Monday, even though it's now Saturday night and you'd rather go to the movies with your friends. You require discipline to save for a new car, as it means not going out every night for a four-course dinner. Discipline is the only way to achieve difficult things and is a hallmark of professionalism.

Soldiers are taught discipline in several ways. The first skills in having discipline are taught at recruit training where all soldiers learn to march on the parade ground and follow orders. They do this regardless of time of day or weather conditions. It teaches them to focus on the task at hand, regardless of the sweat filtering down their brows and into their eyes or the fly that keeps buzzing around their face causing their nose to itch furiously. This prepares soldiers to be disciplined in combat conditions. Discipline is exercised by wearing the uniform (by the setting of standards) so that soldiers still wear their body armour and all the required equipment, even when fighting in ridiculously hot weather where the removal of some uniform items could make soldiers more comfortable. It is exercised in the way their weapon is fired in battle (called 'fire discipline') so that ammunition is not wasted by shooting at shadows or where there is no enemy.

Discipline is a sign of professionalism within a military force, even if that force doesn't have the best equipment or the best training. Discipline can allow individuals and groups to achieve great things against significant odds. During the ISIS rampage into Syria and Iraq in late 2014, the Iraqi army was forced back south towards Baghdad, panicking as a highly disciplined and motivated militia, armed with only basic and often stolen equipment, overran everything and everyone in front of it. At the request of the Iraqi government, coalition forces returned to Iraq, led by US, Australian and

New Zealand training teams, to equip and train the Iraqi army in an attempt to stem the ISIS assault.

I was part of the initial Australian and New Zealand team that commenced training operations, based at the massive former US-run airbase at Taji, north of Baghdad. We had a team of experienced officers and SNCOs, many of whom had previously fought in Iraq in operations during and after the 2003 invasion. Almost all of us were Afghanistan veterans. We had various expectations of the quality of some of the *jundis* (the Arabic name for a soldier) we were going to train. During our deployment we trained new brigade[9] size forces of 700 *jundis* every few months. We often found that some of the battalions we trained within the brigades were very ill-disciplined, often a result of poor leadership by their officers as well as a lack of formal basic training. Unlike western armies, it is common that officer appointments and promotions in the Iraqi army are based on familial or tribal affiliations rather than on merit or experience. This was a source of great frustration for us. We were committed to training the *jundis*, and to see some of them failing to make an effort in their training or simply not turning up to lessons was a test of our own discipline as we had to do everything we could to work with them as that was part of our national strategic commitment to fighting ISIS.

However, despite our own finely tuned professionalism and lengthy combat experiences, we could not help but be inspired by the discipline and professionalism displayed by some of the other battalions that came through our training courses, many of whom had only recently returned from the battlefields in the north of Iraq after having suffered significant losses. These battalions contained *jundis* who, despite poor equipment as well as some poor leadership, displayed levels of discipline in their drills,

---

[9] The Iraqi army was operating units with personnel sizes vastly different from most western nations. The typical size of a western nation's infantry brigade is approximately 2,000 soldiers. The Iraqi army operated brigades with far fewer personnel, one unfortunate reason being an inability to maintain brigades with more soldiers due to losses incurred in the early stages of fighting ISIS.

their weapons handling and in their general approach to their training that we would happily have used as an example of 'what to do' for our own junior soldiers. They were certainly a reminder to us as a western army that we were very lucky to have the equipment supply chains that we did. Most of these *jundis* had to buy or scrounge their own boots and uniforms.

Some of the *jundis* in these units had served in the Saddam era (at a time when the Iraqi army was a vastly more experienced and better trained force) and had reluctantly withdrawn to the relative safety of the south on orders from their army headquarters (HQ). Their discipline was the result of good leadership from the more motivated officers, as well as their individual desire to be professional soldiers, part of the organisation that the Iraqi people were relying on to save the country from the extremists who were slaughtering everyone and everything in their path.

We monitored the brigades we trained once they returned to the fight against ISIS through reports from the frontlines and liaison with Iraqi army HQ. The battalions which we all agreed had shown the most professionalism during our time with them were the ones that spearheaded the successful recapture of the large cities such as Ramadi and Mosul. Their professionalism in battle was a direct reflection of the discipline those *jundis* had.

Having discipline will help develop patience. Patience is critical if you want to be able to make decisions, and more importantly, the right decisions, with respect to your personal and professional life without being influenced by emotion. During peacekeeping operations in East Timor during 2008, as a member of the 3rd Battalion of the Royal Australian Regiment (3RAR),[10] my platoon was often called upon to deal with civil unrest, which often occurred between rival youth gangs. This unrest certainly wasn't aided by a

---

[10] A standard Australian infantry battalion roughly consists of five companies of approximately 100 soldiers each. Four of those companies will be full of combat soldiers (known as rifle companies), while the fifth has all the support staff, such as medics and truck drivers. Three battalions form a brigade. At that time, 3RAR was the Australian army's only conventional paratrooper unit. All Australian airborne infantry operations now reside within special forces.

lack of local law enforcement presence. We would have to stand in between large groups of mostly young men, many armed with rocks, machetes and iron bars. The occasional Molotov cocktail also added to the eclectic mix of weapons. Our job was to keep the groups apart, and often we were armed only with batons and personal protection equipment which included helmets, face shields and body armour.

As highly trained infantrymen, the only corps within the Australian army to have the word 'kill' in its duty statement,[11] it was against all our natural instinct to stay calm and controlled under the goading and attacks by the gangs who quickly forgot about their own rivalries and turned against us. At any moment, as we had the benefit of superior training as well as size, we could have easily attacked and defeated the gangs, seeing as the average Timorese was naturally skinny and well under 6 feet in height. It was our discipline which allowed us to remain focused on our task. For some of our more experienced soldiers, this was especially hard, as many of them had deployed to East Timor during the Australian-led intervention in 1999 after Indonesia had granted an independence vote to the Timorese people. This vote, which was almost unanimously in favour of independence, led to pro-Indonesian militias rampaging in the streets and killing hundreds of pro-independence supporters, often subtly supported by the Indonesian military. An Australian-led military peacekeeping force was sent to restore order, which resulted in Australian soldiers engaging in combat with militia elements and, on the odd occasion, with Indonesian forces. To now be back in East Timor and getting rocks thrown at them was a strange way to be thanked for that effort by some of the local youths. This discipline, learnt over many years of training, ensured that we didn't overreact and potentially cause harm or injury to the civilians, which would have resulted in a diplomatic and international incident as East Timor

---

[11] As per Australian army doctrine, the role of the infantry is to *"seek out and close with the enemy, to kill or capture him, to seize and hold ground, and repel attack, by day or by night, regardless of season, weather or terrain."*

was now a sovereign nation. Such was the trust placed in our training and discipline under the concept of the strategic corporal that we were able to conduct ourselves professionally, which in turn endeared us to the majority of the local population and assisted us in restoring peace to the tiny nation.

Personal discipline is something that can be taught and learned and is a trait that, once mastered, can be applied to anything you do. Discipline allows you to achieve things because, as the saying goes, nothing worth doing is easy. One of the more common teaching methods for instilling discipline currently being spruiked, especially by former military personnel writing books such as this, is to make your bed in the morning. This is a small and simple task which allows you to easily and quickly achieve a goal once you have woken up. It establishes a mindset that you want to take pride in the things you do and do them well, even when no one else is looking or pushing you to do things. Unless you're at recruit school and getting yelled at by the directing staff (DS – essentially the Australian term for drill sergeants), then no one is going to know if you've made your bed in the morning or not. It's a small act that allows you to tell yourself that you've achieved something, and you can commence the rest of your day with a positive attitude and a mindset that you want to do everything that you undertake to the best of your ability and without taking shortcuts. This promotes professionalism and allows you to achieve more.

Another disciplined activity which I promoted in *Battle Rhythm* was the push-up club. This is a fitness-based activity whereby you do a certain number of push-ups each hour over the course of the day, regardless of what you're doing or where you are. Not only is it a healthy activity, but it also allows you to practise making the effort to take the time to undertake an activity that has positive benefits but that could easily be avoided and is reliant upon your own discipline to actually do. If push-ups are something you can't do, they can easily be replaced with a stretch or another physical activity like a walk, both of which are important, especially if you work in an office environment and find yourself behind a computer for the majority of your day. If work gets in the way due to things like meetings, then make the effort to complete the push-ups, or whatever activity you have set yourself, at the first available opportunity. Set yourself a total number you want to achieve each day. This will help instil discipline in yourself.

The Texas A&M Aggies college baseball team instils discipline in its players by making sure that they keep their personal space in the locker room perpetually clean and tidy. The idea is that doing the little things while off the field will be easily replicated when playing on the field. The punishment for any indiscretion is for the player to have to spend some time using the visiting team's locker rooms, which means the player may be isolated rather than enjoying the excellent home facilities. Players learn that discipline in the small things needs to be mastered so that they can have discipline in the large and important things.

Having discipline isn't about being comfortable in what you're doing. It's about still doing it in spite of boredom, fatigue, lack of motivation or any other factor that makes you feel like giving up. By doing these little activities repeatedly, you will begin to view the difficult things in life not as insurmountable obstacles but as challenges that you're able to face and deal with because you have practice in demonstrating discipline. Your discipline will become part of your mindset - a habit and a reflection of your own professionalism. It will, therefore, allow you to approach everything with the belief that you're capable of pushing through difficulties and still achieving your goals even when having to deal with adverse situations.

PERSONAS

A persona is where you exhibit certain behaviours and characteristics which can vary depending on the environment you're in. It's how you speak and interact with people, how you treat them and how you outwardly approach the endeavours you're undertaking. It's your demeanour. The ability to amend your persona appropriately and consistently is a reflection of your professionalism. Having multiple personas doesn't mean having multiple personalities.

A persona is a psychological way in which you adapt to the environment you're in and how you then comport yourself. Soldiers, in the same manner as police, are taught to have a persona which directly reflects the requirements of their job and the expectations of the community they serve. That persona is one of a professional, humble and respectful individual. It's a persona that is aimed at gaining the confidence and support of civilians, so that they trust the soldiers to carry out the duties asked of them. Soldiers will always address civilians as either 'sir' or 'ma'am', as a sign of respect. Even in difficult situations they will maintain this persona because this helps maintain a professional approach to their activities.

Soldiers sometimes require different personas. In line with the concept of the strategic corporal, at one moment they may require aggression and assertiveness to fight in combat with an enemy force and then the next need to be able to demonstrate compassion and restraint when dealing with the civilian population, politely engaging in conversation with an awareness of the local customs. Their professionalism entails knowing what the proper persona is at the appropriate time.

During the large-scale flood in the garrison city of Townsville in Northern Queensland during early 2019, the 3rd Brigade of the Australian army was deployed to assist in the rescue and recovery efforts. The Townsville civilians commented on how well the soldiers conducted themselves and how diligent, polite and respectful they were, even after several days of not sleeping and working in wet and uncomfortable conditions. Many of these soldiers' own homes were underwater, but it was their professionalism that saw them continue their duties despite these personal hardships.

While with 3RAR in East Timor, I often interacted with locals, many of whom had been through very traumatic experiences. My platoon often conducted community engagement events where we helped paint schools or played soccer with children. We maintained a friendly and professional persona consistently, even though playing with demanding kids and trying to appear happy all the time became tedious. However, as soon as a call came on the radio to move to a new location and deal with a developing riot, as frequently occurred, we quickly adjusted our personas. We still remained professional but sought to appear more assertive and authoritative so that we could deal with the rioters in line with our rules of engagement (ROE). Our personas let us show them that we were prepared to use force but would remain disciplined enough to not use it unnecessarily, despite the rioters often goading and threatening us. Once we had dealt with that situation, we had to rapidly transition back to our friendly and happy personas and go back to playing soccer with the kids. Seeing us deal with the troublemakers in a professional and appropriate manner then being friendly and engaging again gained us the support of the local population.

I have served with many officers and soldiers who had diametrically different personas and levels of professionalism. Some were sincere, friendly and engaging, whereas others were – due to lack of a better word – complete arseholes. The ones whom I respected the most, and the ones who were often viewed as being the most professional, were those who could do their job well in difficult situations yet maintain a persona that was always respectful to others and humble in their approach to their duties. They treated everyone with respect and courtesy, regardless of what unit or service they were from. They acted humbly, even if they were from elite units and had far more combat experience than the soldiers they were working with. They were the soldiers who were firm if the situation called for it yet didn't unnecessarily yell at or belittle their subordinates and could remain calm and controlled in all situations. Their personas established the reputation of consummate professionalism that soldiers are renowned for.

A professional persona is about how you speak and interact with people and how you comport yourself when doing your job. This holds true as much for the shop assistant in the earlier example as for the CEO of a large corporate firm. Being considerate of others and not being selfish sets a good foundation for professional and personal behaviour.

Your persona will adapt to whatever it is you're doing. The professional shop assistant is going to interact with customers the same age who are technologically savvy much differently than with senior citizens who are hard of hearing and don't fully understand the new smartphone they are trying to buy so they can call their grandchildren interstate. It's unlikely you'll talk to your work colleagues the same way you talk to your children, as you will likely be more relaxed and less guarded in your family environment. If committed to your gym routine, you'll take a more serious and subdued approach to your session, focusing on the task ahead of you and avoiding distraction, which may come as a shock to those who don't usually see you in that environment and only know you as a fun-loving and extroverted person. The professional part is knowing when each persona is applicable to each environment and situation and being consistent with them so that you maintain the trust and confidence of people around you. Your persona will be a reflection of who you are as a person.

Developing a persona for yourself is essentially about being who you want to be. Do you want to be harsh and authoritarian? Do you want to be kind and compassionate yet firm when you need to be? You can choose your own persona; as I said, it's a reflection of who you are. Just remember that the persona you take on will be how people view you and, in turn, interact with you. If you're all talk yet no action, people will see this, and this will become your reputation. This will be the same for your personal and professional life. Are you a leader who treats subordinates fairly and is respected for being a competent boss? Or do you act too familiar and then wonder why no one does what you ask of them?

If you're a naturally calm and humble individual, then maintain that as your persona. You can still be firm and enforce discipline if you need to or be aggressive if required. If you're not the type of person who is naturally loud and extroverted, then you will struggle to constantly maintain that as a persona. Equally, those around you will question your professionalism because they will think you're not able to control your emotions or will view you as being someone who is unpredictable. The respect from colleagues will come from your ability to do your job in a professional manner. You can still adopt some of the traits that are appropriate to the environment you're in. In the army we train introverts to become assertive and aggressive when they need to be, just as we train the more assertive and aggressive types to be calm and to control their emotions under pressure so that they don't inappropriately react to a rock thrown at their head during a riot and cause an international incident. The use of the bayonet, an outdated weapon of war, is still taught to Australian soldiers so they can undertake the bayonet assault course, a series of obstacles where they learn the appropriate use of controlled aggression.

The ability of soldiers to maintain control of their personas, and to employ the correct ones in the relevant situations, is why the army is so respected. The best advice I have ever received as a soldier was from one of my instructors when I began intelligence training. I had come from the infantry which was an environment where soldiers tend to be loud and deliberately obnoxious, even if they aren't naturally that way, because it helps them to fit in to that type of environment. And so they end up adopting some of these traits in their own personas. Loud and obnoxious isn't my natural persona, and my instructor could sense that. During one brief I was giving while on course, I was being loud and joking around, more out of the habit I had developed in my old unit, I think, as opposed to the requirements of the task at hand. My instructor patiently explained that I wasn't in the infantry now and that I needed to maintain a more pro-fessional, calm and considered persona if I was going to be successful as an intelligence operator. It was brilliant advice. Not long after I had graduated from intelligence school, I found myself briefing special forces soldiers on combat operations, and my persona during those briefs was as important as the information I was giving. Had I briefed them in a loud and obnoxious manner, the commander and soldiers wouldn't have had any confidence in me or the intelligence I was providing, and I would have been sacked – which happened to a few others who hadn't taken my instructor's advice and utilised the correct persona for the situation.

For your own life, both personal and professional, you need to determine and then practise which persona you want to be known for generally and be able to adapt your persona to the environment you're in. For example, if you're in a leadership position and your job is intense and frustrating and you deal with undisciplined people, you will need to have a persona which is assertive and firm. You probably don't want to take that home to your family, so do something that will help you calm down or change focus so that you can adapt to your family friendly persona when you get home and the kids are happy to see you. I have many friends who are considered hard men, very professional and competent soldiers. They are firm and assertive when in uniform. However, when I see them in a family environment, they have a completely different persona altogether, one where they are the dad and not the gruff platoon sergeant. They have learnt to adapt their personas, yet it doesn't change who they are as professional soldiers or as people, or the values that they have. Their persona is consistent to the relevant situation so people have confidence in them. When it seems different, we know that there might be an underlying issue, so we will reach out to support them.

Professionalism is about knowing what persona you need to have, and to maintain it appropriately, for the environment you're in. This holds true for the army and the wider military. Although many of the officers and soldiers in a unit may be the same age and may have served together over several deployments, the officers know they still need to maintain an air of authority as they are the leaders, while the soldiers maintain their respect for their officers because they rely on them to show good leadership. That's not to say that they can't share jokes and can't become lifelong friends – as does occur. I still have occasional beers with several of my former commanders – it's just that there is a time and a place for everything. Maintaining the proper persona comes down to the professionalism of the members of the unit.

In all the deployed environments I have been in, one of the common traits I have noticed among soldiers from many nations is their desire to have fun. The seriousness of military operations and combat is such that, without levity, you could quite easily go crazy. Deployments are filled with long stretches of boredom, peppered by moments of intense action. As Australian soldiers, we have a habit of making fun of each other, called 'taking the piss'. It is a means of jovial interaction which, although seemingly a sign of tension or hostility among soldiers, is in fact a sign of respect and is the way we maintain close bonds with each other. Even among enlisted personnel and officers there is a certain level of jovial banter which occurs, but it's mostly at the platoon and company level. The higher an officer and SNCO rises in rank, the more serious their persona becomes as they are the leaders of the army at the operational and strategic level and they need to maintain a greater air of authority and respect. This banter is often the source of amusement to the US troops who operate with us and who, despite having their own similar ways of bonding with each other, are nonetheless shocked at the intensity with which we do it. The Australian use of the vulgar 'C' word as a source of affection among ourselves is foreign to most of the soldiers, even the US Marines whom I've found to be most like us in their mannerisms. This type of banter doesn't diminish the professionalism of the soldiers because it is always done respectfully and in the appropriate environment. The soldiers will never speak in a derogatory manner to the officers, and likewise the officers will never speak in a derogatory way to their subordinates. A good working relationship at this level is identifiable by the friendly banter among the soldiers, mostly in the form of friendly ribbing about the boss's choice of football team, and never by any personal attacks. When it's time to be serious, the officers and SNCOs take on a firmer persona to allow them to be focused on their job as leaders, while the soldiers will tone down their joviality and follow the instructions given to them with discipline and respect.

Being disrespectful or insubordinate to a superior, or disrespectful to a subordinate or a peer, is a sign of unprofessionalism, as is failing to understand the professional requirements of the position you're in. This sort of behaviour is not tolerated in the army since it undermines the effectiveness and cohesiveness of the unit. Professionals will sort their grievances out collegially, even though there might be some obvious tension. A professional approach is doing what is right and what should be done, even if those above you aren't doing what they should be doing as leaders or supervisors. I once worked on a coalition team on deployment that required me to spend lots of time in the intelligence cell. There was a young intelligence officer who repeatedly flirted with the enlisted personnel, gossiped about other members of the task group and spoke poorly about the superiors behind their backs, often in front of my team. When she was finally berated by a senior officer for her actions, she complained that *he* was being unprofessional for disciplining her! To their credit my team, consisting of some very young soldiers, maintained their professionalism and went on with their duties without getting involved in the gossip, maintaining their respect for her position in spite of her behaviour. Professionalism is as much about maturity as it is about being good at a job; it's about accepting the consequences of actions, seeking to learn from mistakes and moving on.

Acting in a professional manner instils confidence in yourself and also instils confidence in you from the people with whom you interact. Being arrogant, selfish and treating people poorly doesn't make you a professional, even if you're really good at what you do. It simply means you can't control your emotions. You don't need to be a soldier to be a professional. As I mentioned previously, professionalism still comes down to the individual. Being professional is a personal choice, but success cannot be achieved without it.

SUMMARY:

- Professionalism is demonstrating a high level of competence and effort in everything you do and being humble and mature while you do it.
- Setting standards will improve professionalism because it ensures you do things meticulously and develop a high level of proficiency in all that you do.
- Values are the principles which guide a personal belief about what is important.
- Discipline allows you to make the effort to do the things that need to be done to achieve your goals.
- Personas are the reflection of your professionalism and how you treat and interact with people.

Resilience

"Grant me the serenity to accept the things I cannot change, courage to change the things I can, and the wisdom to know the difference."

— Reinhold Niebuhr ("The Serenity Prayer")

Most people have probably seen the fantastic movie *Rocky* with Sylvester Stallone. The story of an amateur pugilist who gets a shot at the world heavyweight title spawned many sequels, one of the most popular being the third in the series which was characterised by the theme song 'Eye of the Tiger' by 80s band Survivor. This song personally resonates with soldiers because many brutal PT sessions have been accompanied by loud music blaring from speakers whose music playlist would almost always include that song. The music is inseparably linked to the movie, not least because it is highly motivating. The reason it's motivating is because it invokes thoughts of the key theme of all the *Rocky* movies – Resilience.

Resilience is very closely aligned to *resolve* in that it's about determination to overcome misfortunes and obstacles, and your ability to do it. It's not only *wanting* to but also being *able* to. It's about getting back up after taking many knocks, just like Rocky did in the movies. Resilience is the ability to endure and recover from hardships, setbacks and even failure, to be able to still get back up on your feet to keep pursuing whatever your task (or mission) is. Any number of things can knock you down; those knocks can be physical or mental. People, animals, the natural world and man-made items all have varying levels of resilience.

Resilience is essentially a by-product of discipline. You can be disciplined and not have resilience, but you can't be resilient and not have discipline.

Discipline is the ability to do what is required as part of a process or to achieve a task, even when things become difficult or the only person pushing you to achieve is yourself and giving up appears to be the easier option. It's usually internal factors that cause you to stop, such as your own lack of willingness to be disciplined (i.e. a lack of resolve). Resilience is where you're disciplined in whatever you're doing but still keep suffering setbacks or hitting obstacles, usually from external factors. A boxer may be doing everything technically right in a fight, keeping their hands up in defence and being patient in waiting for an opportunity to attack. However, a better fighter will still be able to be more dominant. The resilience comes where the boxer (like Rocky) keeps fighting, doesn't quit and looks for another way to land a winning blow. Such boxers have the endurance to keep disciplined and keep fighting. To be resilient means being able to put up with these knocks, endure them and still get on with what you need to do.

The special forces within the army run some of the most arduous and demanding selection courses in the military. These courses, which can be weeks long, are physically demanding and require high levels of discipline from the soldiers seeking to enter the much-revered community. They are pushed to their physical and mental limits, often operating for days at a time in harsh terrain, with minimal sleep and little to no food. However, the DS on these courses are really looking for only one key attribute from the candidates: Resilience. While they still expect demonstrations of physical toughness and mental agility, they want to know if the candidate soldiers have the internal fortitude and ability to push themselves to complete a task despite everything around them seemingly falling apart.

Winston Churchill frequently spoke about the resilience of the British civilian population when it endured consistent bombing by the German Luftwaffe during the early 1940s. Night after night German bombs rained down on the large cities across South East and Central England, causing massive civilian fatalities. Despite all this, the population remained defiant, loyal to their nation (when they might have otherwise begged the government to cede to German demands) and managed to withstand the bombing until eventually the Royal Air Force was able to win the Battle of Britain and control the skies over England.

The Swedish vehicle manufacturer Volvo is renowned for producing cars that people jokingly state are impossible to break and keep running for years after other cars have been retired. The way the cars are built and the high standards of design, testing and construction are why this brand builds among the most resilient of all cars. These cars have the ability to be driven for many years, even if they have covered many kilometres on different types of roads, and are able to withstand serious crashes, protecting the occupants and often being able to operate again with only minor repairs.

Resilience is important in the army because of the nature of soldiering. The demands placed on soldiers, and their equipment, far exceed those found in any other profession. Combat does not adhere to set times or plans. Soldiers are required to deal with any situation that they may face and endure the hardships of their operating environment. This may involve having to operate in austere environments for months at a time, with diminishing supplies and in the face of a relentless enemy. There is an outstanding documentary titled *Restrepo* which perfectly highlights the resilience of soldiers. The film follows the fortunes of 2nd Platoon, Bravo Company, 2nd Battalion of the 503rd Infantry Regiment of the US Army during their 15-month tour of the Korangal Valley in eastern Afghanistan commencing in May 2007. During their tour, the goal of which was to clear the valley of ACMs as well as gain the trust of the local population, the men of 2nd Platoon faced adversity almost daily, ranging from having to build an outpost in some of the harshest conditions on the planet to regular firefights against insurgents, which occasionally resulted in the loss of a fellow soldier. Despite all these hardships and setbacks, they were able to build and maintain operations of the outpost and made some headway into gaining local population support by the time they handed over to the incoming company. Their resilience was forged not only from their training but also from their commitment to each other as brothers-in-arms and an inherent desire to not let each other down.

The latter stages of the Australian commitment to Afghanistan, before withdrawal from main combat operations in Uruzgan province in late 2013, involved training and partnering with Afghan National Army (ANA) soldiers in operations against the ACMs. In several separate attacks over the last few years of the commitment in Uruzgan, several rogue ANA soldiers conducted what is called a green-on-blue[12] attack on their Australian counterparts, resulting in seven soldiers killed and more wounded. Several coalition forces experienced these green-on-blue attacks during their commitments in Afghanistan. To endure those significant setbacks yet continue to train their ANA counterparts, as well as to conduct patrols[13]in open plains and villages littered with IEDs,[14] all the while at risk of being shot by someone who was supposed to be ally, was a testament to the discipline and resilience of all soldiers who served in Afghanistan, regardless of which country they came from.

The personnel of both the US and Australian armies are well disciplined; it's part of who they are as a professional force. However, one of the questions often asked by defence analysts and strategic planners is *does the force have resilience?* By this they mean not only the personnel but also the equipment and the procedures that are used. New technology is fantastic, and in the modern age many of the army's communications items are highly digitised. However, it's no good having a fantastic computer that can speak to a satellite to guide munitions to an enemy camp if the computer can't operate in dusty conditions in the mountains of Afghanistan where the infantry needs to use it. It's pointless having an attack helicopter fleet that can't support amphibious troops landing on an island because the aircraft

---

[12] In military terminology, 'blue' forces are considered all those under the coalition banner, whereas 'green' forces are those local forces who are coalition allies but under indigenous command arrangements.

[13] A patrol is where infantrymen walk or take a vehicle to a defined area, usually to deter enemy activity or to determine an enemy presence. A patrol can be made up of as few as four men or as many as several hundred.

[14] Improvised Explosive Devices – essentially homemade bombs

can't operate over water. These are examples of a lack of resilience in the equipment, meaning the equipment cannot deal with external influences and keep working well.

The level of resilience is important in that the more resilient something or someone is, the more that piece of equipment or person will be able to do. Military commanders seek to have the best equipment available to conduct operations, knowing that the equipment will be subject to the rigours of warfare and will need to operate in multiple environments. They seek to have tanks that can operate well in the mud and can still operate even if they have been hit by a few rounds of an enemy machine gun or rocket system. They want their soldiers to have clothing that doesn't fall apart after a few months of exposure to swamps or humid environments.

Resilience is built into soldiers and their processes through their training. It becomes part of their mindset and an expectation of their job performance. When on exercise, the training scenarios always simulate the worst possible conditions that might realistically be experienced so that soldiers become conditioned to having to deal with setback after setback while still striving to achieve the mission. The army deliberately makes the soldiers experience the 'worst-case scenario' in any training activity so that their mindset is attuned to dealing with the setbacks which may occur. For example, the commander or the platoon sergeant in an infantry platoon might be notionally 'killed off', which means the next soldiers in line have to step up into the leadership positions. The resupply truck will notionally 'break down', which means the soldiers will have to get used to extending their rations and water. The transport helicopter will mysteriously be called away to another mission, which means the infantry platoon now has to walk the 20 kilometres over undulating terrain to the next position ready to conduct an assault on an enemy position. There was a line often quoted by one of my corporals when I was a private soldier in the infantry about going on exercise and operations, which holds true for any infantry unit around the world: "Expect to be cold, expect to be tired, expect to be hungry… and expect to do it all walking." However, this sort of training naturally builds a depth of character in soldiers and increases their natural resilience. They have the ability to accept and endure difficult situations and continual setbacks yet still focus on getting on with the job and doing it well.

During my first deployment to East Timor in 2007, my company deployed with less than eight hours' notice, and unlike other deployments where we had the opportunity to take a foot trunk with some of the comforts of life, we deployed with our packs only, which were pre-packed and ready for this type of rapid deployment. To our chagrin as paratroopers, we didn't jump into East Timor but landed by Hercules aircraft instead at Dili[15] airport. We were immediately deployed to the steep and highly dense jungle hills in the centre of the small island nation, ostensibly to assist our special forces to capture a militia leader who had been part of a failed coup attempt the previous year and was now causing trouble for the East Timorese government from the relative safety of the countryside.

We were operating in true infantry style, living off whatever we could place into our packs, leading patrols of several weeks' duration in what seemed like never-ending rain and conducting risky operations aimed at seeking out the militia leader and his armed followers. A few weeks into our deployment, the first of what became a flood of 'Dear John' letters started arriving for the men in the company. This was a time before smartphones and Facebook were ubiquitous, and old-fashioned letter writing was still the best (and in this case the only) form of communication. The letters were received by soldiers in various stages of relationships, ranging from girls some of the guys had met at the local sports club only a few weeks earlier to a few instances where engagements were called off. It was like something out of a movie, and once we were moved back into the capital a few months later, some of the members of Pioneer Platoon[16] kindly built a noticeboard to which these letters were pinned for all to read. To every

---

[15] Dili – the capital city of East Timor
[16] Pioneers in an infantry battalion specialise in basic combat engineering, such as digging fortified trenches and setting up booby-traps.

single man's credit, not one soldier asked to be returned home, and all chose to stay with their mates and see out the deployment.

On one particular patrol, my platoon suffered numerous non-battle casualties, all a result of the mountainous terrain, the constantly inclement weather or the illnesses acquired in tropical environments. We lost our platoon commander in the first minute of the patrol, even before insertion into the operating area, when he rolled his ankle while walking towards the Blackhawk helicopter on the improvised helipad that was usually a soccer field. Over the next 11 days, we lost more soldiers for various reasons. One of our machine gunners nearly fell off the side of a mountain, saved only by a platoon mate who dived at him with an outstretched hand at the last moment. Others fell ill due to the tropical environment. I personally did several rolls down a hill while fully laden with my pack which, I'm told, while impressive was hardly at an Olympic gymnast standard. Each injured soldier needed to be medivaced[17] out, either by helicopter or by our Land Rovers which would meet us at a rendezvous point somewhere near the patrol route. We became affectionately known within the battalion as 'the platoon that went on patrol that returned as a section'. Despite setbacks such as these, we kept focused on our mission, which involved several months of heavy workloads without knowing when we would be relieved by another unit. Our resilience was strong despite its being tested constantly. We didn't even experience any combat on that deployment. If you throw in a few firefights and IED strikes, you'll start to understand the experiences of the many infantry units who served in Iraq and Afghanistan.

The army has proven itself resilient, particularly over the past 20 years. Multiple deployments by soldiers, often using outdated equipment which has been patched together with spare parts and made to work due to the ingenuity of the soldiers using it, has imprinted resilience as a mindset and part of the culture in the army, as well as in its soldiers. It is one of the reasons soldiers are highly sought after in the civilian workplace as well as

---

[17] Abbreviation for 'Medical Evacuation'

by professional sporting teams and governments, all seeking to understand how the resilience instilled in and demonstrated by soldiers can be used in their own specific organisations.

Resilience is an enabler that will greatly enhance your ability to achieve success in your personal and professional life because nothing ever comes easy. Resilience is the tool you use when the challenges in your life start to feel overwhelming, and discipline is being called upon repeatedly. Why do I say resilience is so important? Because life can sometimes suck. Like combat, life is unpredictable and often doesn't happen as you'd like it to or as you plan for. Just because something should happen one way doesn't mean that it does. It's the very epitome of Murphy's Law.[18] As such, you need to be adaptable to whatever life throws at you so that you can still do the things you want to do and achieve your goals, despite all the external things that can, and often do, get in the way. The analogy I like to use, especially when teaching soldiers about the reality versus the theory of warfare, is that combat is like a zebra crossing. In theory, you should be able to walk onto a zebra crossing to go across to the other side of the road. You *should* be safe in the knowledge that the crossing is adequately marked and has flashing lights and that you're protected by the road rule which states cars are supposed to give way. The *reality* is that, even with all those things in place, a drugged-up driver distracted by a mobile phone can still come flying through the crossing at 100 kilometres an hour, with the potential to kill anyone in his or her path. You just have to be prepared to accept that reality, manoeuvre around the car and be cautious when going over the crossing, doing what is needed to ensure you make it over to the other side of the road safely. Once you accept and adopt the mindset that that is what life is sometimes like, you will have taken the first step to becoming resilient because when the bad things do happen and your efforts seem to be thwarted at every turn, you won't be

---

[18] The notion of Murphy's Law is that anything bad that can happen in relation to something or to a situation probably will.

inclined to give up. Instead you'll get back on your feet and keep working towards your goals.

Life will always test resilience. The loss of employment, a sudden health issue or having your house burn down are times that require strong resilience. While these are traumatic events, having resilience will allow you to deal with the situation with a clear mind and help you avoid making the situation worse by making emotionally charged decisions or giving up on things all together.

Most children are taught by their parents at an early age that simply whining about a situation, such as falling off a slippery slide or dropping an ice cream on the ground, doesn't change or fix the situation. They're taught that they have two options: to sit there and keep crying or to get up, brush themselves off and get back up on the slide again (or hope they might be lucky enough to salvage what is left of the ice cream). The most resilient children are the ones who keep falling off their bikes, who keep struggling at school or who keep playing on a team in the local sporting competition that gets beaten every week, yet still get up, keep trying and have another go. These lessons are learned at a young age and are taken into adult life.

There is a narrative in some parts of the media and wider society these days that millennials lack resilience. It suggests that they're part of a generation that has spent its youth inside playing video games and has never experienced some of the hard knocks of life, and as such is not prepared for the 'real world'. I find that to be true in some cases, but very unfair in most of the other ones. I have served with and instructed junior soldiers, some still in their teens, who are highly resilient, motivated to serve and eager to learn. They lack the sense of entitlement that I have found while working with some of their peers in the civilian sector. A lack of resilience is not confined to a certain age group, though. I have seen many examples of people of various ages who throw the towel in after getting knocked back from only a few job interviews or who commence gym routines and quit after only a few sessions.

You can instil resilience in yourself and use it as a means of not giving up when things become difficult in your life or you find yourself struggling to achieve the things you want even when you're ostensibly doing everything required. Ensuring you have instilled discipline in yourself is the first step. Without discipline, you will never have resilience because you won't be used to doing the little things properly and under pressure, which means you won't suddenly be able to do them when a higher and more consistent level of effort is required. Endeavouring to take a professional approach in your life is the second step as you'll view the difficult things that are getting in the way of your goals as challenges rather than obstacles. You'll be better conditioned to take a considered approach to how you will deal with problems and seek various solutions to them as opposed to giving up easily. Making an honest effort to learn from your mistakes, attempting to understand where you went wrong and what you can do better the next time and realising how you can approach the situation differently is all part of this professionalism. You won't waste time by doing the same thing repeatedly. The most successful and resilient corporate companies and sports teams are the ones that critically assess themselves and identify where mistakes have been made and where mistakes could be made. Doing this ensures they survive, and thrive, during economic downturns or when the best players are injured.

As soldiers are tested in exercises to build resilience, you can also do similar things to help build resilience in yourself. Building physical resilience means getting used to enduring hardships and training the body to withstand heavy burdens, without going too far and causing avoidable injury. It takes discipline to get up in the morning and do PT when you're still tired and it's cold outside. Training in shorts and a singlet will build resilience because your mind gets used to dealing with a bad situation (being cold and outside doing push-ups) when you would rather be in a better situation (being warm and in bed). Taking a cold shower will build resilience in a similar manner. It's cold and uncomfortable, but it's not going to hurt you. Doing this will condition your mind to deal with the situation so that when you start to suffer difficulties or setbacks in your life, your mindset is such that your focus is on dealing with the situation rather than throwing in the towel.

Building and maintaining strong relationships enables resilience because you will have support to assist you when things become difficult. Knowing you have people who encourage, support and assist you will improve your ability to endure setbacks and hardships. Soldiers develop incredibly close bonds, forged in combat and from their shared military experiences, which most people cannot really understand because it's not something replicated in the civilian world (perhaps the closest situation is a very tight sporting team or a firefighting team). These bonds last a lifetime. Soldiers will fight and serve for many reasons, but when the bullets start flying, it's all about looking after each other and doing everything they can to all get home safely. You don't have to have a large friendship network. As I mentioned in the chapter 'Centre of Gravity', choosing your friends carefully is essential to ensuring that you can rely on them when needed.

Doing these things will give you a strong centre of gravity as they build the confidence, self-belief and sense of purpose which you'll need to face adversity in your life. When external factors make the challenges look insurmountable, this confidence will stay strong if you remember to ask yourself *what am I trying to achieve?* and *how badly do I want something?* It's a simple thought process that allows you to recognise that you need to show some resilience in order to keep moving in the right direction towards whatever it is you want to achieve. Failing at something, or doing something not as well as you would like, doesn't mean you lack resilience. Resilience doesn't eliminate stress or pressure. It's about overcoming adversity, going back to have another go and thinking about ways you could do it better when you do fail or have bad days. Success can never be achieved if you quit.

At some point, you may have to make a decision that something you're pursuing is not achievable or simply no longer worth the effort despite your best discipline and your resilience. Some things in life are simply insurmountable. (Think of the plane getting shot down. It was never going to be able to defeat a missile, even though it's a resilient piece of equipment.) Sometimes, it's no longer worth the effort. In 1863, during the US civil war, the Union forces attempted twice to capture a large Confederate battery at Fort Wagner on Charleston Harbour. Two large assaults during July by Union soldiers were repelled by the Confederates, who were then besieged and had to endure constant bombardment by Union artillery for several months. The Confederate soldiers showed great resilience by constantly enduring so much pressure and a mounting casualty rate. The fort was never captured by Union soldiers. Instead it was abandoned later that year when the Confederates no longer deemed it a vital military asset. It wasn't a lack of resilience that forced the soldiers to leave but rather a change in their military strategy.

There is a threshold for resilience for everything. Even the most hardened and experienced soldiers have a threshold for their own resilience, and the intensity of the numerous deployments undertaken in the past 20 years has taken its toll on many fine soldiers. They are only human after all. The suicide rate amongst Australian veterans is far higher than that of the general population, and the US is suffering similar numbers proportional to the rest of society. It's a tragedy which unfortunately continues, despite ongoing efforts to address the situation. Attempting to enhance your own resilience is going to improve your ability to deal with the challenges that life throws at you and will increase your chances of succeeding in achieving your goals. However, always keep in mind that everyone has a threshold, so never, ever be afraid to ask for help when you find things are becoming extremely difficult. Soldiers are part of a team, and teams always help each other.

ATTITUDE – 'EMBRACE THE SUCK'

Resilience is as much of a mindset as it is a tangible quality. The human body is amazingly resilient. Special forces tests for resilience by putting the candidate soldiers through some amazingly demanding physical testing.

The candidates are free to withdraw at any time of their own volition. A few will be medically injured and literally cannot go on, while the vast majority who quit do so because they mentally cannot handle the stress anymore. Their bodies are tired, they are exhausted and hungry, but it's their mental doubts that make them want to stop. Their bodies are quite capable of continuing the demands of the training, but mentally they cannot go on. It's resilience that makes the successful soldiers complete the incredibly difficult training and go on to lead storied careers, often in the shadows and in some of the most dangerous situations imaginable.

Often the ability to endure such hardship comes down to attitude. There are three types of attitude: positive, negative and neutral. A positive attitude is taking a 'glass half full' approach to something; a negative attitude is taking a 'glass half empty' approach to something, whereas a neutral attitude essentially means you have no firm views or opinions on something. The attitude you take towards doing things has a direct relation to how well you're able to deal with the difficult situations that might happen while doing it. How you view the situation also impacts how you're able to deal with it. Attitude is an inherent part of resilience.

The soldiers who are the most successful, whether during special forces selection, on peacekeeping operations or during combat, are those who have a positive attitude. A positive attitude is about undertaking endeavours with the intent to do a task as well as possible, despite any hardships experienced while doing those tasks. Doing things any other way results in failure because mentally it means they've given up. The army uses psychological warfare operations as a means of undermining the morale of its adversaries by instilling a negative attitude in them to make them believe that their situation is dire even if they are actually in a good position and still have the majority of their resources intact.

Cyberspace is quickly becoming a new domain for the conduct of warfare. The use of social media during the fight against ISIS was an effective means of undermining the morale, and thus changing the attitudes, of many of the fighters who had taken up the terrorist ideology. The army engaged in non-kinetic warfare, through the use of social media posts, to undermine the personal beliefs of many of the ISIS fighters who were attempting to wage their own form of warfare on the internet by posting gruesome pictures of their atrocities. However, these pictures served only to harden the resolve of many nations to defeat the terrorists. The attitudes I saw among many of the Iraqi *jundis*, despite having experienced defeat, having lost comrades and being expected to fight, often with substandard equipment, was both humbling and inspiring. They were fighting for the survival of their country and their families and for their lives, and so their attitudes were based on an unshakable desire to achieve victory.

Soldiers of the US army have a slang term, long embraced by other armies, which they have used for decades and which is often quoted in television shows and movies. It not only sums up a predicament (being in a crappy situation) but is essentially an unofficial mantra for how they conduct operations. It's called 'embrace the suck'. In other words, they fully accept and appreciate that something is unpleasant but also realise that it's unavoidable. They are fully cognisant of the problems that a certain situation presents yet still remain optimistic that a positive result can be achieved. Therefore, they just have to accept it and get on with the job as best they can.

Being in a warzone in combat or undergoing training in a harsh environment is never enjoyable. However, if it's the situation soldiers find themselves in, which in reality is by choice considering the modern army is an all-volunteer force, then instead of viewing the situation negatively and thinking about how miserable they are, they choose to embrace it and get on with the job. It's still unpleasant and uncomfortable, but it needs to be done. There's a line of dialogue in the movie titled *The Odd Angry Shot* (about a group of Australian soldiers serving in the Vietnam war) that most modern Australian soldiers refer to that is similar to the 'embrace the suck' mantra. When discussing their choices in life and how they ended up fighting in Vietnam, one of the lead characters says to his colleague, "Everyone has to be somewhere, and you're here ... so get used to it!"

A commander will always want to deploy with soldiers who have the best attitudes and will heavily rely on the officers to present an accurate update on the morale of the soldiers within a unit. Morale doesn't necessarily mean enjoying or being particularly enthused about the prevailing situation. Morale is about the level of confidence, discipline and resilience among a group at any particular time; morale also includes maintaining a belief in the objectives of the mission. Good morale is achieved through having a positive attitude.

Soldiers take the proven approach that the best way to get out of a bad situation is to simply get on with the job, so that they can either complete the difficult training or finish the course or exercise they're on or get home from the deployment safely. They rely on their discipline, resilience and the support of the soldiers around them to achieve this. Embracing these situations, along with other soldiers, forges the camaraderie that the army is renowned for. It allows soldiers to achieve the things they need to do because it prevents an enemy from being able to undermine their morale.

At most Australian army bases, there is an obstacle course. It's made up of a series of physically challenging obstacles, of varying difficulty, that are designed to replicate the hazards of operating in both rural and urban environments. The obstacle course is one way that the army builds resilience, as well as physical fitness, in its soldiers. Regardless of the size or structure of these courses across the bases, one of the first few obstacles is always 'the bear pit'. This obstacle is essentially an open pit filled with water, deep enough to come up to waist height on an average-sized person. Soldiers have to jump into the pit, with their camouflage uniforms, weapon and webbing, and wade through to the end of the pit which slopes up back onto the ground. This means that they have to complete the rest of the course in heavily soaked clothes which increase the weight they are now carrying – difficult at the best of times. All soldiers know they will face this drenching, and so they just get on with it.

During one of my deployments to Afghanistan, I worked within the HQ element doing intelligence duties on a large Australian-run coalition base in the middle of the country. Due to the sensitive nature of the communications equipment within the HQ area, a gated fence was built surrounding the internal compound (often referred to as 'a compound within a compound'). Security was important as the Australian campaign was being run from these facilities, and there were local civilians working on the base, mostly as cleaners and cooks, who were not permitted access to this area. A problem developed when the gates to the internal compound were not being shut properly by the personnel passing through them as the spring mounts (which were designed to make the doors shut by themselves) were broken. The issue was known to all personnel who were working inside the compound, and a command directive was released instructing everyone to ensure the door was being shut behind them (which was actually part of usual security protocol anyway). Despite the command detective, it was still noted that the gates were being left open, placing the security of our sensitive equipment and personnel at risk.

I was asked by the head intelligence officer to find an inconspicuous place near the fence line and spend a few hours observing who the biggest culprits not shutting the gate were. I took some notes, and even a few covert video images, to verify my findings. It turned out that it was the Australian officers who were the biggest offenders and followed proper security practices the least, followed closely by the US officers. I delivered my findings to the intelligence officer, who subsequently delivered them to the commander. About a day later, the decision was made that a guard would be placed at the gates, made up of the junior soldiers (not officers!), to ensure the gates were being closed. I couldn't believe it. I spoke to my team, as well as to some of my counterparts whose soldiers would also make up the guard, all of whom were about as impressed as I was. Most of my team were already working 16-hour days, so to add an extra two hours watching a gate because of the failings of certain elements of our leadership was a hard ask. I wrote up a roster and personally took the first and last shift of each day (the longest shifts as it was a 24-hour roster). I gathered all the soldiers who were to be part of the guard, both US and Australian, briefed them on their duties and emphasised the concept of embracing the suck. I couldn't have been prouder of all the soldiers, who not only watched the gate with eternal vigilance but who also enthusiastically began to act as doormen, opening and closing the gate for each officer who passed through, warmly greeting them and wishing them a good day. This was designed to show that we could handle anything asked of us, no matter how stupid, and also served to subtly send a message to the officers who had caused the situation. After a few days, the commander reversed his decision about the guard (after the doors were finally fixed) but not before, as I later discovered, several of the better officers had approached him and politely demanded that their soldiers not be utilised in such a manner. We just embraced the situation and went on with our duties, as good soldiers always do.

Being able to develop a positive attitude regardless of the situation you're in is how you can be successful in your personal and professional endeavours. You can do this by reminding yourself about the bigger picture and telling yourself why it is you've set out to do something in the first place. Remind yourself of the motivations behind beginning an endeavour. Children are taught that procrastinating in cleaning their room won't magically make the task go away. They soon learn that the best way to achieve the task, and to be able to get outside to play with their friends, is to just accept the situation and get the job done. Procrastination is one of the biggest reasons why people don't achieve their goals, whether it's because they don't go to the gym when they're supposed to or because they're not feeling very motivated to write the essay for university.

Embrace the suck in all you do, and you'll find that the things you thought were painful or difficult suddenly become just small obstacles, easily overcome, that you need to get through in order to get to your goals. Nothing will be achieved if you sit back and hope that a situation is going to get better. Those 100 push-ups won't do themselves. Just remind yourself why you're doing them and what the benefits will be once you've completed the relatively small physical activity. Attending the mandatory training day at work may be as boring an activity as you can imagine. View it as an easy way to get paid, and consider that the break from the office routine may be cathartic. Take it as an opportunity to meet other members of your office whom you haven't met yet and to develop your networks, which may make all your subsequent work efforts easier. Maintain your self-discipline and get on with the job at hand. I'll speak in a later chapter about how to view things from different perspectives and how this can influence your attitudes towards the things you do in your life so that you can better adapt to difficult situations and maintain your resilience.

Developing and maintaining resilience will help build confidence in your own ability to overcome difficulty when you fail, if all appears lost or if the odds against you seem overwhelming. This will, in turn, increase your self-belief and your ability to overcome these difficulties that are getting in the way of you achieving your goals. Resilience allows you to get back up and continue pursuing your endeavours. Resilience will become a tangible part of your character and abilities and make you keep going until you achieve success!

SUMMARY:
- **Resilience is getting back on your feet when you get knocked down. It's about having the endurance to keep doing what needs to be done so you can achieve your goals.**
- **Accept that life is often not fair and that it's up to you to adapt to its dynamics.**
- **Resilience leads to a strong centre of gravity, which builds confidence in your own abilities to achieve your goals.**
- **Attitude is about the way you approach all your endeavours. It's a way of thinking and feeling about something.**
- **Embrace the suck!**

Dealing With Conflict

"The clever combatant imposes his will on the enemy,
but does not allow the enemy's will to be imposed on him."

— SUN TZU

A testament to a person's professionalism and resilience is how he or she deals with conflict. The very purpose of the army is to engage in armed conflict by confronting potential enemies at the behest of the nation. Soldiers are highly trained to use a myriad of weapons to conduct warfare in any environment on the planet. Conflict, whether physical (kinetic) altercations or tense arguments and disagreements between two or more parties, is unfortunately a reality of life and an all too familiar part of the shaping of the history of humanity. Conflict can take many forms, from two children having a verbal argument in the schoolyard to nations sending their militaries to conduct armed warfare against other nations or large entities such as ISIS.

Discussing how to deal with kinetic conflict is to take on a subject that has many theories and hard learned lessons. Many academic books have been written on the subject. I will confine this discussion to conflicts described as non-physical experiences that occur in personal and professional environments. These are usually arguments, robust discussions and disagreements that can occur between work colleagues, family members or complete strangers.

Soldiers are as well trained in conflict management and conflict avoidance as they are in conducting conflict (i.e. combat). Soldiers have been called upon more and more over the past 20 years to be involved in non-warlike operations – for example, peacekeeping and peace-enforcing, as well as humanitarian support operations – which can often see them interacting amongst potentially warring groups. This is why the army trains soldiers to manage conflict using non-lethal force, because the very nature of conflict, especially one where weapons are utilised, implies that it almost always results in casualties. The nature of the mission an army might be assigned to, such as peacekeeping, means that casualties amongst civilians must be avoided at all cost. As a result, even though soldiers have the skills and equipment to engage and easily defeat an adversary, the potential cost to neutral parties is too high, and so unit commanders will do everything they can to avoid actual kinetic engagement. Instead, the soldiers will seek to de-escalate or rapidly resolve a situation that may be evolving, such as a riot in East Timor, so that people don't get hurt by a potentially lethal over-reaction to that situation.

Soldiers are taught discipline so that when they find themselves conducting riot control duties (or other non-lethal operations where the use of lethal force isn't expected to be required) and getting rocks thrown at them, they don't escalate the situation unnecessarily to a level that might involve lethal force or result in unacceptably high levels of violence and injuries. That's not to say that they just stand there and take it, which would be a hard ask of even the most disciplined soldier. Because they are equipped with the necessary tools to protect themselves and are trained to gradually escalate the amount of force they use, only offenders receive the wrath rather than innocent civilians.

The presence of 'curious onlookers' is something I have experienced in all theatres I have deployed in, from East Timor to Afghanistan. It's simply the nature of peacekeeping and asymmetric warfare: there will always be bad guys living and hiding amongst the civilian population. Therefore, in any operations we conducted, we always had to take into consideration the potential effects on the local population. When a non-lethal situation requires higher escalation, soldiers will use the skills and equipment available to them. Using non-lethal weapons, such as tear gas, is one method used by security forces globally to disperse crowds and calm down a situation, without people being unnecessarily injured.

The idea is that conflict, when it occurs or looks like it's about to occur, can be managed. The whole purpose of each nation's diplomatic corps is to promote its national interests and to avoid conflicts with others. The army is usually brought in only when these attempts to avoid conflict have failed or look like they might be about to fail. Bringing in the army is often a last-minute effort to try and avoid any further escalation by deterring the other side through a display of military power.

Nations such as Australia and the US have a proud history of assisting smaller nations in times of trouble, either to help avoid conflict before it happens or to assist when it is already occurring. The US formed a large military coalition in the Middle East after Saddam Hussein invaded Kuwait in 1990. The initial strategy was to try and scare the Iraqis into leaving Kuwait and retreating back across the borders. The large military build-up was an attempt to avoid war. Unfortunately, Saddam didn't believe the coalition seriously intended to fight, and a short but deadly war soon followed. In 1999, Indonesia permitted one of its small provinces, known as East Timor, to vote for independence. The former Portuguese colony was ethnically different from Indonesia who had invaded it in 1975, killing tens of thousands of Timorese in the years that followed. Over 90% of the population of East Timor voted for independence, which sparked mass violence in the streets as Indonesian-backed militias went on murderous rampages in response to the vote. Australia responded by leading a United Nations mandated military force (the International Force East Timor – INTERFET) into East Timor to quell the violence and to monitor the Indonesian withdrawal. There were a few skirmishes between INTERFET and the Indonesian military, as well as numerous clashes with the militia.

This is why soldiers are trained in conflict management. They are taught techniques to deal with conflict and confrontational situations and become comfortable working in those environments (part of the concept of the strategic corporal).

Few people relish conflict or confrontational situations, for example, being berated by the boss, getting yelled at by a fellow driver for a perceived slight on the drive home or being the victim of an armed robbery at home or in the community. Conflict almost always results in the body naturally having an involuntary chemical reaction (known as the 'chemical cocktail') either in anticipation of or in response to the conflict. We all react differently to this; it's the 'fight or flight' response. Some people react positively and can deal with the situation calmly – mainly because they've been trained to deal with such situations or have experienced the same situation before and have learnt what their own reactions are and how to deal with them. Others begin to shake uncontrollably. Their field of vision narrows (i.e. they lose awareness of the environment around them), and they might have trouble formulating a coherent sentence. This is a perfectly natural reaction, too; it's nothing to be ashamed of.

Even the most experienced soldiers still feel that initial moment of terror and shock when the first crack of a bullet passing over their head is heard. The speed of a bullet fired from an automatic weapon is so great that a distinctive noise, which is the bullet literally smashing through the air faster than the speed of sound, is heard. This means that soldiers hear the bullet pass overhead before they hear the sound of the rifle that fired it. This is commonly referred to by soldiers as the 'crack and thump'. The conditioning that comes from repetitive scenario-based training is why soldiers react with a 'fight' response. When they are faced with the real thing, they have learnt what their natural response is and how to deal with it. Soldiers are also taught calming techniques to help bring down their rapid heartbeats when the adrenalin starts to wear off. Deep breathing to a three-second cadence whilst simultaneously moving their fingers and toes – to help get the blood back to the peripheries of the body to prevent potentially going into shock – and looking around slowly to increase the field of vision is a proven technique when starting to feel fear. It works, so consider using it when you get that nervous feeling in the pit of your stomach the next time you have to go and speak to the boss unexpectedly! Although it's unlikely you'll ever face the crack and thump that soldiers do in war, you can still use some of the skills that they're taught in order to deal with conflicts in your own life.

Most of the conflicts that people face in their personal and professional lives are with people they already know, whether they be friends, family or, in many cases, work colleagues. The conflicts are usually verbal disagreements about issues which are genuinely serious or are escalations of some smaller original disagreement. However, these confrontations usually turn out to be emotional reactions devoid of logic and fact. Often the two parties have 'bottled up' their thoughts and feelings until a point is reached where one decides enough is enough and finally confronts the other person. In the workplace or in social environments such as sporting clubs, conflict typically stems from issues relating to things like missed promotions, disagreements about ways to do things, how people are being treated and spoken to, or something as simple as who is making a mess in the communal kitchen. Most workplaces these days have some form of procedure for dealing with employee issues, and the success of these processes usually depends on the abilities of the leadership in that workplace to promote and enforce them.

Talking to your boss to seek a pay rise or having to deal with a troublesome employee if you're a supervisor can be an unnerving experience. Such confrontational situations may lead to conflict, especially if the other party doesn't respond favourably. The same applies to arguments with a friend or family member or trying to get a stubborn and incompetent shop assistant to give a refund for a damaged item you have purchased. These can be unnerving because there is the risk, whether it's perceived or otherwise, that an argument or tense conversation may break out or that you may not get what you're seeking. Don't be afraid of this!

Here are several tips you can consider if you think you're about to be in conflict or are already in conflict with someone. This is not a procedural guide but simply some of the techniques soldiers use. The first tip is: *Nip it in the bud*. If you find yourself in a position where you might have to deal with a potential conflict, do what you can to resolve it early or to avoid it all together. Soldiers use cunning, bluff or force overmatch to scare and defeat an adversary. During the invasion of Iraq in 2003, elements of the Australian Special Air Service Regiment (SASR) were conducting operations in the western part of Iraq. Part of their wider mission was to capture the large Iraqi air force base at Al-Asad. Because they had already been in numerous engagements with enemy forces that were superior in number, the shooters[19] were keen to avoid more of what was an already very high enemy casualty rate. The SASR commander initially used a loudhailer to tell the Iraqi soldiers held up in the airbase that the war was effectively over and that they should give up. When that failed, he directed a Royal Australian Air Force F/A-18 Hornet to conduct a low-level flyover of the base. The sound of a fighter jet breaking the sound barrier at 100 feet above the ground was intimidating enough to force the remaining enemy troops to surrender, and no further engagements were needed. Questioning of the Iraqi soldiers after the event suggested they were scared that going outside when first petitioned by the Australians on the loudhailer would have resulted in a firefight, but they were more scared that a bomb would be dropped on them after hearing the aircraft flyover. Smart thinking by the soldiers resolved the conflict.

Seek to avoid conflict early by asking yourself whether the issue can be resolved another way or by determining early on what all the causes of the potential conflict are and addressing them with the appropriate people. Most of the conflicts we have in our lives are usually because people don't communicate enough in the early stages, don't bother to find out the concerns of each party involved or don't seek another approach to the situation. Don't hold grudges because they will distract you from whatever it is you're trying to achieve and can have long lasting negative effects. One of the biggest issues we found in trying to gain the support of the population during operations in Afghanistan was that some villages maintained grudges against other villages based on real or perceived slights, which in

---

[19] Common term for qualified members of special forces units such as the Green Berets, Commandos and Navy SEALS

many cases went back hundreds of years. As we found out the hard way in the early part of operations, this meant that we had to be very careful about how we used information provided by villagers about dispositions of ACM in the area. For example, some elders of one village told us that a second village up the road was full of ACM. Based on this information, on more than one occasion, coalition forces initiated an assault on the second village which sometimes unfortunately resulted in civilian casualties. On further examination after the assault, we discovered that there was, in fact, no ACM in the second village and that the information that had been provided to us by the village elders was deliberately false. We were being lied to so that we would inadvertently kill or detain the first village's opponents.

Use whatever processes are available to you to avoid conflict. Speak to your boss if you have a problem with an issue or a colleague at work and are having no success resolving it yourself. Ask to speak to the store manager if you're not getting the service you'd like from the shop assistant. Suggest an alternative to the current situation if someone is refusing to budge on any particular issue. Sometimes you just have to walk away and avoid the conflict altogether even if you really believe you're in the right. Personal relationships are areas where early resolution of an issue will avoid potentially damaging the relationship in the long term.

Following on from that is the second tip in relation to conflict: *Don't sweat the small stuff.* This is a well-known mantra that is often taught to children but one that many people seem to forget when they enter adulthood. The rapidly evolving pace of technology in modern times means people are trying to do more and fit more into their lives without any change to the time they have available to do it. This inherently increases stress levels, and as a result, people often react poorly to little grievances.

While training to deploy to East Timor with 3RAR, we conducted an activity called force-preparation training whereby another specifically designed unit in the army conducted mock exercises to simulate all the scenarios we expected to face while on deployment. This was to condition us to deal with any fight or flight scenarios, although in peacekeeping the 'fight' often meant simply keeping focused on our mission and not over-reacting to situations, such as a riot, that could result in adverse strategic implications. Soldiers from other companies within the battalion were role players, acting the part of civilians involved in a riot. Our job was to protect a mock 'vital asset'[20] from the 'rioters' who were attempting to get in. The vital asset was actually the battalion transport yard which, in this instance, was standing in for an East Timorese government building.

Our orders were to use non-lethal force to prevent the rioters from entering the asset. Non-lethal force is still force nonetheless and can involve the use of tear gas and batons as well as physically detaining the worst troublemakers in an attempt to deter the rest of the mob. Once the rioters, who were mostly our good mates, realised that they wouldn't be able to get past us and into the building, they changed their focus to us specifically and made nuisances of themselves in order to annoy us as much as possible. We began to receive the most horrible forms of personal abuse and had improvised projectiles thrown at us. Some of these consisted of old rotten food, balloons full of urine and the odd full soft drink can, even though the latter had been specifically prohibited by the DS due to safety concerns. The paratroopers of 3RAR, who made a living jumping out of perfectly good aircraft, often had a different view of what constituted 'safety' from the rest of the army. This type of training conditioned us to not take things personally and to not react unnecessarily to these attacks as we had other things to focus on – our mission. As long as no one was getting past us, we were achieving that mission. We had to stay disciplined enough to deal with things in the appropriate manner. When we eventually deployed and had to deal with actual riots and general population unrest in East Timor, we did so in a professional manner. When the time came to escalate the use of force, we did so and inflicted swift but measured violence on those whom we needed to control, but always in accordance with the ROE and always

---

[20] A 'vital asset' is typically a piece of infrastructure that the well-being of the community depends on, such as a power station.

as a means to protect lives. Many other times we simply had to stand there and take the abuse, and the odd projectile, because that was our mission, even though at any point – as highly trained and motivated paratroopers – we could have destroyed the people attacking us. We learnt to not take it personally and to take the attitude that we were the bigger men.[21]

The same should apply in your own life. Don't allow yourself to take offence at things that don't matter or to get frustrated with things that can be dealt with in a different and more appropriate way. That's not to say you should just accept bullying and abuse, not at all. That sort of behaviour is not acceptable in modern society, and most workplaces and institutions have processes in place to deal with that sort of thing. If those processes are available, use them appropriately. However, you don't benefit from wasting your time on the small things you can't control or that happen infrequently. A person cutting you off in traffic is annoying, but hitting the horn doesn't achieve anything except giving you a moment of satisfaction that could end up escalating the situation unnecessarily. You need to consider what your task (your 'mission') actually is, which could be to get your kids to school safely and on time. Professionally, you may be working with a colleague who never pulls his or her weight or is never on time and lies to the boss. This might not affect you directly, but it's certainly very frustrating nonetheless, especially if you're a consummate professional. Don't get frustrated by this. Use the processes available to you, such as informing the boss (who is ultimately responsible for dealing with staff), as opposed to taking matters into your own hands and potentially saying or doing something that could get you fired. Often, simply removing yourself from the situation gives you clarity, as frustrating events that occur in a sheltered environment can appear to be bigger than they really are. Keep your focus on your mission, which is doing your job as best you can.

---

[21] Up until 2013, females were not permitted to join the combat units (designated as infantry, armour and artillery units) of the Australian army, so on this deployment all the qualified infantrymen were male.

The third tip in relation to conflict is: *Be polite, show respect and get the facts.* Soldiers are renowned for always being courteous and respectful. Soldiers are always polite and respectful to the people they encounter, regardless of whether they're in their home country working on a base or on patrol in a dusty village in Afghanistan. It's part of their professionalism and instils confidence in the people they're working with. It also has operational advantages. As I stated earlier, the identified centre of gravity for the war in Afghanistan was the ability to have the support of the civilian population. Therefore, every effort was made to always be polite and courteous to the people who lived in the villages that were being patrolled, even if there was intelligence to suggest that the people supported or harboured ACMs. Soldiers took an ample supply of lollies and chocolates to give out to the children, and the head of the patrol ensured that appropriate compliments were paid to the village elders, usually in the form of the sharing of chai.[22]

Law enforcement officials are taught the same thing. They are polite and courteous to civilians, even when on the receiving end of shocking abuse from a drugged-out criminal or an irate driver whom they have rightly pulled over for speeding. Showing someone some simple respect can help defuse a situation and avoid conflict, as will succinctly stating your point of view or opinion. However, if that fails, continue to be polite even if the person you're dealing with continues to act belligerently. You do this by maintaining your professionalism and not taking things that are said in the heat of the moment personally. There is a great scene in the movie *Roadhouse* where Patrick Swayze's character is hired to bring a higher level of professionalism to the security staff in a dingy saloon. He begins by firing the most ill-tempered and corrupt staff, then reinforces to those remaining that their job is to not take verbal abuse personally and to remain polite even to those they are escorting out of the door by force, explaining why they are doing so. This technique ensures that emotion is never brought into the conflict and allows the staff to maintain their professionalism.

---

[22]   Chai is a form of tea popular in South Asia and the Middle East.

Being polite may be enough for the person or people you're having a disagreement with to rethink their position. They may pause to consider that they escalated the issue unnecessarily, and they now realise that you're an amicable person who is willing to listen to their perspective and come to a resolution. If not, then in the world of modern technology where there are CCTV and smartphone cameras seemingly everywhere, how you've comported yourself can be recorded, and that may be enough for the people who come in to resolve the issue (whether the boss in the workplace or someone with authority such as a judge) to take your side or to make a decision in your favour. Police officers will tell you that when they arrive at the scene of a car crash or a domestic incident, the person they're most likely to listen to first or give the benefit of the doubt to is the person who has his or her emotions in check and is being the most courteous to the officers.

In relationships, you should always be striving to treat your loved ones with respect and courtesy. At the end of the day, it's family and close friends who will be there for you if you really need them, but they're less likely to want to be in your life if you treat them poorly. Soldiers commonly refer to their fellow soldiers as brothers-in-arms (and sisters-in-arms. Warfare doesn't discriminate against gender and many females have performed superbly under fire as soldiers.). This is because of the camaraderie that comes with living and working with a group of people in some of the harshest and most dangerous environments imaginable. Soldiers need to be able to trust their fellow soldiers implicitly, so any issues between them get resolved quickly to ensure that, when they're in combat, there is no lingering resentment that could affect team cohesion. Soldiers come from the wider society, and there will always be those who don't get along. However, even when those who don't like each other work in the same team, they will still respect each other based on the fact they have each undertaken the same training and are in the same place, perhaps fighting daily in difficult combat situations. Without that respect, there is no team unity, and an enemy will easily exploit this weakness in battle.

In 3RAR we had different ways to quickly resolve internal conflicts between soldiers. Some platoons made arguing soldiers rehearse and give a tactics or weapons lesson together so that the soldiers had the opportunity to learn a bit more about each other and develop respect between them even if they were never going to be best friends. In other platoons, it wasn't uncommon for the platoon sergeant to get the boxing gloves out, take the soldiers out to the parachute training hanger (a concealed area) and let the soldiers go at each other until there was a distinct winner or until those soldiers decided that their grievances against each other weren't that serious. At the end of the fight they were made to shake hands, and if there was still serious animosity between them after that, the officers would step in and take administrative action, removing the soldiers from the company, battalion or even the army if required.

While I don't encourage fighting to resolve issues, I do encourage making the effort to deal with conflict by using politeness, respect and courtesy. If that doesn't work, look to the processes that are available (e.g. escalate to a higher authority) to deal with the conflict.

If you get the opportunity, watch a senate estimates committee (in Australia) or a senate armed services committee (in the US) hearing where select politicians, from all sides of politics, ask questions and seek clarification about issues regarding the military, whether related to defence spending or any other issues at the politicians' discretion. At all times, the senior officers respond to the questions in a courteous and respectful manner, maintaining the professionalism that the military is known for, even when a politician is being rude, disrespectful, deliberately hostile or, in many cases, simply incompetent or tries to make himself or herself look important at the respondent's expense. The officers avoid conflict by not engaging with these politicians in an argument or being rude in response to their hostile questions. That's not to say that they let the politician walk all over them. On the contrary, they will certainly respond assertively if needed and defend their respective service as required, but they will always do so politely and respectfully. They do this because these hearings are subject to public scrutiny, and the officers want civilians to see firsthand the professionalism of the military. Civilians are more likely to trust and be confident in their military once they see this (even if they don't have the same confidence in and respect for their politicians!).

When having a disagreement with someone, you can remove emotion by finding out the facts of the situation. Always take a moment to assess the situation that is in front of you so that you understand well what courses of action may be available to you to deal with the conflict. Soldiers who are on patrol in combat zones are highly trained in the drills used to respond to the crack and thump. The first step always involves assessing the situation and making a decision about what they need to do to respond. This process takes only a few moments. The required response may be for the soldiers to fire back at an enemy with as much gusto as possible. Sometimes the response is to hold fire and try and determine where the shooting is coming from and ensure that it isn't from another friendly force that mistook the patrol for an enemy, or from an elderly farmer who has protected his land for decades using an old rifle and who also mistook the patrol for a potential intruder. The professionalism of the soldiers ensures that they don't choose the former option when the latter one is correct.

Additionally, you should make clear what your opinion or point of view is or why you're having an issue with something in the first place The best police officers will keep calm and explain their actions to someone they're arresting or giving a speeding ticket to, even in the face of hostility and always showing respect. You can do the same when dealing with situations of potential conflict, such as an issue with a colleague. Have your facts ready and, calmly and politely, state your point of view and be prepared with some solutions to the problem. Most people want to avoid conflict, especially in the workplace, because they know they have to work with the same people consistently, and a harmonious workplace is a happy workplace. If you're in the right on an issue, outlining the facts to someone articulately and politely will make them more likely to accept that they are wrong. They will be more receptive to finding a solution if they feel that you're not trying to hurt them, personally or professionally.

The fourth tip in relation to conflict is: *Pick your battles.* Soldiers engage in combat under one of two circumstances: either as a deliberate assault on an enemy position, or in response to an attack such as an ambush. A deliberate assault means an intelligence-led, well-planned and repeatedly rehearsed attempt to capture a bridge, a fuel depot or, as occurred frequently in Iraq and Afghanistan, high value individuals such as the leaders of the ACMs or Saddam Hussein and his loyal followers. Doing this is called 'maintaining the initiative', which means the attacking force is essentially controlling the battle. When responding to an attack, soldiers have to instantly react to the situation, following well-practised and well-rehearsed drills in order to take cover and to begin engaging the enemy. Concurrently, the commander seeks as much information as possible to determine the size of the enemy, its capabilities (i.e. weapons and tactics) as well as its locations so that a decision can be made whether to conduct a counter-attack or to make a fighting withdrawal and get to safety. (As the saying goes, "He who fights and runs away, lives to fight another day.") A decision like this may be made for several reasons. During operations in Afghanistan, the ACM hid in the villages and often attacked approaching patrols from the small houses (known as 'qualas') in the village. Despite often having superior weapons and training, soldiers within these patrols could not counter-attack if there was a risk that doing so would harm or injure the civilians who were almost always still in the village. In our ROE, both Australian and US, any kinetic action that could unnecessarily risk the lives of the civilian population was prohibited. In these instances, patrols withdrew from the engagement area and approached at another time and from another area. This was very frustrating, but the soldiers were disciplined enough to deal with the situation and understand the requirements of the ROE. It was far better than having to live with the knowledge that preventable civilian deaths had occurred.

In both of these situations, the commander always needs to decide if engaging the enemy is worth the effort – whether the benefit is worth the potential cost, which may be the loss of soldiers' lives. In the case of an attempt to capture a high value individual, the result may mean a loss of leadership for the ACM, which could cause it to falter and, in turn, stop all attacks in an area. In a responsive situation, a unit may need to engage the enemy because it's the only way to get out of the engagement zone, and their very lives may depend on being able to fight their way to safety. One of the first recipients of a gallantry medal during the early stages of the Australian involvement in Afghanistan was a member of 2nd Commando Regiment, a senior soldier who maintained his highly exposed position in the lead vehicle, staying behind the machine gun and returning fire as the patrol sought to fight their way out of an enemy ambush.

You should pick your battles in your personal and professional life. It can be very easy to waste a lot of time in an argument or disagreement for which there is very little benefit and where the cost to you could be high. In a professional scenario, for example in a corporate environment, is it worth continually arguing with the boss over a point of contention that may not really matter in the overall scheme of things? You may have to critically ask yourself if you're too invested in something simply because you've spent a lot of time and effort on it, and if you want to pursue a course of action (which the boss has now said no to) simply because you don't want to think you've wasted your time and effort on something that is not being used. If that ends up happening repeatedly, then it's probably worth having a chat with the boss about whether you're being utilised in the workplace correctly. However, not taking the direction of the boss or arguing with colleagues on moot points wastes your time and their time and could result in a loss of confidence in you, which may affect your chances at the next promotion.

The best army commanders never operate in isolation. They have their principal staff officers, such as their intelligence and operations staff, on hand to advise them at all times. Good commanders always listen to their staff and receive input. Great commanders give their staff as much time as is available to try to persuade them to follow a certain course of action, then explain the decision that has been made and get the soldiers to believe in it to undertake the mission. As a principal staff officer working as an intelligence operator within special forces, it was my role to inform the commander (colloquially known as 'the boss') as to the enemy situation and to help inform his decisions as to the best action he could take. Sometimes my information was readily taken on board and significantly informed his decisions, while on other occasions he decided to overlook my input or take a course of action I personally didn't agree with. In these latter instances, I had to be more assertive (politely, of course) with my input to get the boss to understand that I believed that the intelligence I was providing should be given more weight in his decision-making process. My challenge was to decide which points were worth being more assertive about and which ones were probably not worth arguing about. If I argued all of the time, especially if I ended up being frequently incorrect, his confidence in me would have diminished. The benefit of picking my battles was that, when I did become more assertive, the boss would usually give me another chance to explain my point of view as he understood that I didn't disagree with him unless I had a firm belief in my opinion. Once he made the decision, I followed it as if it were my own, and he respected me for that.

In your personal life, whether in your relationships or when interacting with others in society, you must also choose what is worth having a conflict over and what is worth taking issue with. You must look at what the cost and benefit to yourself is with regards to taking issue with something. If you've had your car serviced and the wheels fall off the moment you drive out of the auto repair shop, then this is certainly something you'd want to talk to the mechanic about and take issue with, because you now don't have a car that works. The cost and benefit analysis about taking issue with this is easy, as you've paid good money with the expectation that the mechanic will provide a service that is vital to your ability to get to work, to drop kids off to school or to be able to carry your groceries.

Conversely, a pizza shop forgetting to place a certain topping on your pizza is really not worth making an issue about, especially if you've driven all the way home and the cost of fuel to go back and complain is worth more than the extra topping. Is it really a big deal? It may feel like a big deal if it happens over and over again. Well then, ask yourself why, if this keeps occurring, do you keep going back to that store? If you know that they keep forgetting a topping, ask yourself what you have done to improve the situation. Did you make an effort to tell them that you expressly wanted a certain topping? Is going back and complaining immaturely to the manager really setting a good example for your children? Perhaps going back to the shop and speaking to the manager, in a polite and respectful way, is the perfect opportunity to set a good example for your children, and your approach will demonstrate to them that issues can be dealt with if done in the appropriate way. You might even get a free pizza from the apologetic manager. The decision has to be yours, but always consider the cost and benefit analysis first.

As I said when talking about resilience, the world isn't a fair place a lot of the time. Many things happen that are plainly unfair. What should happen isn't always what does happen. There are plenty of times where walking away from something is the right thing or the best thing to do, yet doing this can be very frustrating. Deal with this frustration by ensuring you have made an honest assessment of the cost and benefit of your decision to not pick the battle and by applying a professional approach to that decision. In other words, did you consider the bigger picture, or are you sweating the small stuff? You'll be surprised how easily you can take solace from knowing you've made the right choice, knowing that you've saved time and effort from being wasted unnecessarily or realising that you've avoided some potentially bad consequences. Picking your battles is not about allowing yourself to be walked over by people, it's about ensuring that you're not wasting time and effort on issues or matters that aren't worth it. It can be a fine line, and often understanding when to fight comes from hard won experience.

Dealing with conflict is about getting the resolution you're seeking, which in most cases is to end the conflict or resolve the confrontation without being worse off for it. You have to keep emotion out of it by focusing on the bigger picture (i.e. your mission) so that unnecessary distractions don't get in the way of whatever it is you're trying to achieve. When you start to get that feeling that comes with the chemical cocktail, you need to take a deep breath and take a moment to consider using some of the suggestions listed here. Unfortunately, conflict is a part of life, and there is no formula for avoiding it altogether, apart from staying home and never going out into society. But you won't ever achieve anything this way. You need to recognise that conflict happens, and once you understand how you respond to it and find what tips work best for you to deal with and resolve it, you will be better positioned to pursue your goals and be more successful in your endeavours.

SUMMARY:

- **Tips for dealing with conflict:**
  - Nip it in the bud.
  - Don't sweat the small stuff.
  - Be polite, show respect and get the facts.
  - Pick your battles.
- **Focus on your tasks (i.e. your mission, the big picture) as a technique to avoid allowing emotion to get the better of you when dealing with conflict.**

Think Smart, Act Smart

"There is always one more thing you can do to increase your odds of success."

— Lieutenant General Hal Moore

Soldiers are taught a certain way of thinking, which influences how they act because they have to consider a wide range of factors and possibilities when performing their duties. The concept of the strategic corporal means that soldiers need to be highly trained, physically and intellectually, so they are equipped to deal with the myriad of scenarios that they could be presented with, whether in barracks, on peacekeeping operations or in combat. The level of responsibility they are often entrusted with requires maturity, insightfulness and an acceptance that at times they may have the lives of other soldiers in their hands. This is because soldiers are required to make decisions, often with little time available to make them, which could have fatal consequences. Soldiers don't take on this burden lightly, and it's the experiences they gain while bearing these burdens that moulds them into the highly skilled professionals whose talents are eagerly sought in the civilian world.

I've spoken about what a centre of gravity is and how leveraging your own helps to develop your personal and professional life. I then discussed how soldiers are professional and resilient, and how these attributes can be personally developed so that you can approach your own life with a more confident and considered manner, able to adequately deal with conflicts that may occur.

Throughout their careers soldiers learn skills, formally and informally, that allow them to think critically and to consider all perspectives when making decisions. This is so they can conduct their duties in a professional manner, and it gives their commanders confidence that the men and women they send out to conduct operations are equipped with the ability to think on their feet and will not make preventable mistakes that could have dire consequences. The army doesn't want Rambo-like characters who shoot from the hip and seemingly kill everything in their path. While there is certainly a need for courage, controlled aggression and the ability to follow orders on the battlefield, modern soldiers need to be able to think independently and make decisions on their own. This wasn't expected of the soldiers of previous generations who conducted warfare. That's not to say that the soldiers who bravely fought the wars of the past didn't have the need to think for themselves – far from it. While soldiers have always been renowned for their ingenuity and creativity, it's simply that the nature of modern warfare, often without front lines and in the glare of the 24/7 news cycle, means that soldiers have many more things they need to think about when they conduct their duties. They need to think about how their actions will affect the mission and how they can do their part to ensure its success without making avoidable mistakes which could have negative strategic consequences.

While the army teaches many of these types of skills to enable the soldiers' success on the battlefield, I will discuss a few key ones that I think will best assist you in your own personal and professional development as they are easily transferable to almost any environment. They are skills that you can add to your own 'toolbox' of concepts that will enable success in your own endeavours.

FAILURE IS NOT AN END STATE – YOU NEED A PLAN B

On joining the army, I believed myself to already have a solid level of fitness. Unfortunately, *army fit* and *civilian fit* are two different things, as the physical requirements of soldiering entail more than going for a scenic jog or doing a set of push-ups. The first time my platoon at recruit school did the obstacle course, we all failed the mandated time. Being young, motivated men, we all took the perceived failure fairly personally. I remember sitting around as a group at the end of our first attempt at the course, exhausted and wondering what punishment we were going to receive (and wondering how we were ever going to beat the required time). Our platoon sergeant, a short, stocky and very gruff infantryman by trade, walked over and eyed our dishevelled group of trainee soldiers. "What the f*** is wrong with you lot?" he hollered at us. Being the oldest of the platoon, I was usually expected to answer for my teammates. "We failed, sergeant," I replied. He eyeballed me, looked back at the rest of the platoon and started laughing. "So what!" he said. "Do it again, but next time go faster," he added matter-of-factly before walking off. Once it dawned on us that we weren't about to be kicked out of the army for failing one test, we continued to face all our tests and challenges with the same motivations and vigour but without the fear that failing would mean the end of our careers. We knew that we simply had to work harder and try again. At the end of the recruit course, my section set the fastest time for the graduating class.

In combat, failure can result in casualties and can risk the success of the overall mission. The Dieppe raid is a good example of a failed military mission. Good soldiers died despite their courage, skills and resilience. Many mistakes were made. However, the worst thing that could have happened as a result of this would have been to not have learnt any lessons from the failure. The failed raid did not lose the war. The mistakes were reviewed, the lessons applied and the D-Day landings at Normandy were successful.

I have spoken at length about resilience and the ability to continue your endeavours despite facing hardship or difficulty. Often, it is possible to have resilience, to keep going and continue your efforts and still fail anyway. There are many things in life that, despite our best efforts, may simply be beyond our reach. I will never, no matter how hard I try, be good at mathematics (the difficult stuff like algebra. I can calculate the change from a twenty after buying a beer.). I studied hard at high school yet repeatedly failed anyway. I demonstrated resilience. I tried again and again. I still failed. As a result, I accepted that mathematics was never going to be my forte, and so my dreams of being a fighter pilot were dashed. (I failed the entrance test due to poor mathematics and exhausted the permitted number of retests.)

I needed a Plan B.

My failure was not going to be the end of my pursuit of a career. I simply had to look at other options. Failure was not going to be an end state for me. I ended up pursuing a career in the navy – which I stuck with until I decided to join the army instead.

Like my mathematics efforts, you can attempt things repeatedly and still not make the grade. You've shown resilience; you've shown the desire to chase that goal. Thinking and acting smart is knowing when it's time to move on to something else, when to approach the goal from a different angle or when a particular objective or goal is probably just not within your reach or still worth the effort. The key is knowing the difference between accepting that it's not achievable and quitting. Quitting *is* an end state.

 Most people will never be able to run the 100-metre dash in under 10 seconds. Only the elite few ever will. Others will try as hard as they can, devote all their time and get the best training, but it's just not achievable for all. However, for the majority of physically abled people, running a marathon is achievable. There is no time limit; it's about covering the distance. Most people quit mentally before their body actually gives out and before they really find out whether they are physically capable of running a marathon. With good training and a willingness to make all the necessary sacrifices, it is possible to achieve this highly regarded feat. Even a failure on a first, second or subsequent attempt can be overcome. Quitting will ensure the goal is never achieved. Perseverance, however, will either help achieve the goal or expose previously unknown limitations that make it physically impossible to actually run the 42.2 kilometres consecutively. At this point a decision has to be made that perhaps this particular goal is not achievable. Wouldn't you rather know for sure than die wondering? From here you can look to set other goals and undertake efforts to achieve them. You pursue your plan B.

Special forces candidates who undertake the selection course of various units will often be given a chance to undertake the course again if they fail initially. Failure to complete the course based on a self-withdrawal is, except for rare the exception, usually cause to end an aspirant's chances to serve in those units. Those candidates who fail for medical or physical reasons, who are still literally dragging themselves through the mud or up a hill to complete a task despite an injury, will often be given second chances as they have not quit and have shown the resilience that is highly sought by those units. After further attempts, they will either be successful or have to unfortunately accept that their bodies are not suited to the rigours of working in such a demanding environment. Regardless of the cause of the failure, these soldiers still have a career in the army, and many go on to serve and have distinguished careers in other areas. While they take it personally for a while and begin to have self-doubts about their abilities, colleagues and superiors remind them that they still have plenty to offer the army and they need to look back at their own centre of gravity to remember where their strengths lie.

One of my best friends, whom I joined the army with, once failed the infantry reconnaissance course early in his career. This is possibly the most difficult course an infantryman can do, next to the sniper course, and usually helps to identify candidates who will eventually be suitable to undertake special forces selection, though that particular course is not a pre-requisite. He took this first failure not as reflection of his character or capabilities but rather as an identification of his shortcomings, which meant he was now better prepared to pursue his goal of serving in special forces. His positive attitude was the difference between quitting and getting up to try again. He successfully went on to serve in special forces in the 2nd Commando Regiment, one of Australia's most highly regarded units.

Failure in life is not like failure in combat. With a few exceptions, such as surgeons and airline pilots, your failures will not result in people dying. Your failures are not a reflection of your entire life. You will have the opportunity to try again or to choose to go on to a different pursuit. If you can accept that it is possible that you might fail in an endeavour or a pursuit but realise that it will not be the end of the world if you do, then you will become less risk averse and more willing to try things. This in turn will open up opportunities to you that you may not have previously been aware of, and you will certainly learn more about yourself. You cannot achieve success by not giving something a go. Failure allows you to become better at your endeavours if you learn from your mistakes. Failure is an opportunity to try again and to do better. Failure gives you the hindsight you didn't have the first time around. It's about having the right attitude. Demonstrate resilience by failing and then trying again. Fail again then try a third time. Always learn from your mistakes and attempt to work out what you need to do better or differently so that you can be successful next time.

However, if you have given something your best effort and you've made the logical decision that a particular goal or pursuit is simply not achievable, then have your Plan B ready. For example, when I wanted to be a fighter pilot, I had to accept the fact that it might not happen due to reasons that were beyond my control. (In my case, I couldn't meet the academic requirements.) I still wanted to serve my country and found satisfaction doing so by jumping out of planes instead of flying them.

NEVER STOP LEARNING

A soldier's career is characterised by lengthy periods of training and, if the geo-political situation dictates, shorter periods of time on deployment and potentially on combat operations. The periods at home or away on exercises and undertaking courses ensure that soldiers are conditioned and trained to the highest possible level so that they can face any adversary in any environment. Training is demanding and closely replicates, within safety parameters, the actual combat conditions.

At the end of every activity, an after-action review (AAR), which is an assessment of the successes and failings of the activity, is conducted so that lessons can be learnt and applied. The army is always learning. It understands that physical skills atrophy, so that is why soldiers spend hours on the range firing weapons. This practice ensures their skills become highly tuned so that they maintain an expertise in the use of their personal weapon. Physical conditioning deteriorates, so that is why soldiers conduct physical training and conditioning, whether in the desert heat or the mountain cold. As a result, they are ready to deploy in shape and are able to deal with the harsh conditions combat presents. Tactics and equipment change, so that is why soldiers undertake courses and study the latest developing trends. Thus they are appropriately prepared to face any adversary on the modern battlefield.

Learning, and keeping yourself informed about the world around you, should be a lifelong activity. It should become a habit. Thinking that you know everything and that you can't possibly learn any new skills or improve your knowledge will only result in mediocrity. It is an arrogant attitude and, unfortunately, one that sometimes occurs in some of the more senior soldiers (officers and NCOs alike) in the army who rest on some very weak laurels. The best soldiers are the ones who are eager not only to perfect existing skills but also to learn new ones that will assist them in their ability to perform their duties. They are the ones who seek to increase their knowledge and are open to new experiences. Modern military technology is advancing at a rapid rate. There are always new weapon platforms that need to be learnt and understood, along with their appropriate applications in battle. Doctrine will frequently change as a result, so new tactics need to be developed and taught. Therefore, it is incumbent on soldiers to keep themselves informed of all these changes so they can maintain their proficiency.

Never be above learning from those who may be subordinate to you or junior to you in experience. The higher that soldiers ascend in the army, the less often they are 'on the tools' (i.e. using the equipment and training in the field), as they inevitably go into administration and management positions. This means that the most proficient soldiers in tactics and weapons are the junior soldiers, as they are out training every day with the equipment and have learnt the latest tactics and skills. Therefore, when the senior soldiers require refresher training, it is the junior soldiers who give the lessons. The army, despite all the modern technology available, still maintains ongoing training in basic skills to ensure those skills remain current and available if needed. These skills include land navigation using only a compass and a map, even when GPS is available, or determining distance to a target for snipers, using landmarks and natural terrain to judge the distance, even when laser range finders are available.

I spent over a decade studying ju-jitsu and other martial arts under one of Australia's most prominent and well-respected instructors, *Hanshi* Rob Gear. A former military policeman, he rose to the rank of 10th Dan[23] and ran his own suite of training academies. It was a very old school way of learning, as he had studied under some of the pioneers of Australian martial arts, such as multiple world record holder *Hanshi* Bruce Haynes. It was brutal training and, in all honesty, probably as hard if not harder than any of the training I had undertaken in the army. By the time I earned my black belt, which was by no means on the first attempt, I felt like I could achieve anything I set my mind to – at least once the bruises had healed! However, one of the key lessons he always taught to us was humility. Early in my training, a former student came in to do some training. He

---

[23] Different styles of martial arts have colour belts to indicate achievement of a certain rank. Black belt traditionally signifies recognition of having achieved a high proficiency in physical and academic training and testing. The ranks above this are called degrees, or 'Dans' and are achieved through specialist training and by becoming proficient in instructing. A 10th Dan is essentially recognition of a lifetime of training and commitment.

was a 5th Dan and a highly proficient martial artist but hadn't trained for many months due to work commitments. He asked me to give him some refresher training and help him polish his techniques. It was nerve-racking but very humbling at the same time, as it taught me that you're never too good to learn from others even if they're technically subordinate or less experienced. We always strived to become exponents of our own style but to be open enough to learn from others as well. We attended numerous seminars and brought in guest instructors and students from other dojos, all with the goal of sharing information and knowledge. If another group of students was training in a certain way or doing certain techniques that we thought were good, regardless of their experiences or rankings, then we would incorporate them into our own training. It was a lesson all his students took with them into their other endeavours.

When I served in a special forces unit, some of the qualified special forces operators occasionally came into the intelligence cell where my colleague and I, both ex-infantrymen, worked. We gave them advice on how to best set up their packs and equipment for training and living for extended periods in the bush. They sought our assistance because, in the mid-2000s, the Australian army implemented a concept whereby soldiers could be recruited directly into special forces from the civilian world, as opposed to having to spend several years in the conventional army first as had usually occurred. The Intelligence Corps initiated a similar concept. Both these trades sought additional means of recruitment in order to help meet the demands of continual service in the War on Terror. Although some of the soldiers in special forces and intelligence units were highly trained and experienced in asymmetric and counter-insurgency warfare, they hadn't gained the experience and skills that came from years of training in the Australian bush using doctrinal methods for living and operating for weeks at a time using only what could be carried on their back in their large packs. This became a problem when it came time for these soldiers to do promotion courses which were conducted by units specifically tasked with training soldiers in preparation for higher rank. These training units were within the conventional army, and all trainees were required to conduct the field operations phase using strict conventional doctrine, including the carriage and use of field equipment.

Australian soldiers, based on experiences in the jungles of New Guinea during the Second World War and then again in Vietnam, had developed highly refined jungle warfare doctrine which, by the end of the Cold War, was regarded as the world's best practice. Even with some of the best jungle warfare training and experiences in the world, we were not above learning from the East Timorese soldiers who we were helping to build their fledgling military. Many of these soldiers were former guerrilla fighters who had spent years running a resistance campaign against the Indonesian occupation, and so their jungle warfare skills were highly effective and, more importantly, battle proven. We taught them how to run a disciplined military, and they taught us jungle warfare skills that helped us update and improve our own doctrine.

One of the hallmarks of both Australian and US special forces soldiers is their desire to be the best at all that they do. When we deployed together, it didn't matter that we didn't wear the same beret. What mattered was that we were able to help them become better at operating in the bush, just as they patiently helped us become better at packing our equipment and taught us the skills for conducting helicopter assaults and other special forces specific tasks we were assigned to support.[24]

Modern soldiers, in line with the concept of the strategic corporal, are expected to develop a wider range of skills and capabilities than their predecessors. Some become linguists, others gain advanced medical skills, and some become qualified personal trainers and adult-learning instructors. This is part of the reason soldiers are highly valued in the civilian workforce.

---

[24] Qualified Australian special forces soldiers are awarded a coloured beret to mark completion of their selection and reinforcement training – a Sherwood Green Beret for Commando qualified soldiers and a Sandy Beret for soldiers entering the Special Air Service Regiment.

Seeking to expand your own horizons by learning new skills and gaining new knowledge will not only keep your mind active and allow you to expand your networks but also improve your employability and allow you to become more confident in yourself. Learn first aid so you may save the life of a loved one. Volunteer in your local community so you can meet new people and contribute to society. Try a cooking class so you can improve your ability to cook nutritious meals in line with your gym routine. Read widely about things or listen to podcasts so that you can become competent in the skills and knowledge required by your chosen profession or in areas of interest to you. Attend seminars or join professional associations so that you can constantly learn from others and share your knowledge and skills with them. Even in everyday life you can learn something, whether passively or actively. For instance, if you catch a taxi regularly, engage the drivers and ask about their day and their experiences. Some drivers have amazingly interesting stories that you might learn something from.

Early in my career, I was in a taxi in Sydney with a submariner friend. We had had a few drinks and were trying to get to the other side of town. We started chatting with the driver, a diminutive Nepalese man who spoke English reasonably well and was friendly to us and happy to chat. We told him about how busy our respective weeks had been – mine out field with the army and my friend's at sea underwater. We quietly thought we were pretty hot, being military men after all! We asked him how his week had been, expecting a fairly tame response in comparison to our own. He politely replied that he had enjoyed his week, but then curiously added that he was familiar with spending long hours under the stars doing military training and was happy to now just be driving taxis. Upon further conversation, it turned out that we were being driven around by a decorated former Ghurkha who had significantly more service and combat experience than both me and my friend combined. Our lesson was not only one of humility, but he also gave me some great advice about looking after leg injuries after long pack marches, which Ghurkhas are highly proficient in as they are shorter in height than most western soldiers and have learnt how to deal with the injuries associated with carrying heavy packs over long distances.

Being constantly open to new ideas and new experiences will build your toolbox of skills and knowledge. Learning constantly improves your ability to be successful in your endeavours. Learning helps to keep you disciplined and motivated because it prevents you from feeling that you're not progressing towards achieving your goals. Just as you make time in your battle rhythm to train your body, you should also make time each day to train your mind. Read every day, whether it be a book on a topic of personal interest or a journal article about your profession. Read newspapers so that you can keep abreast of current affairs. All soldiers maintain situational awareness of the environment they're working in. It allows them to be as informed as possible so that they can make the right decisions. Do the same so you can be situationally aware of the world around you, so you can be well informed about things that may impact you, your life, your endeavours or even your loved ones. It's not always just the morbid bad news that you should focus on. Most newspapers have articles on finance and health that could be of interest or help you to learn something new. If possible, keep a pen and paper with you so you can write things down that you think are valuable. Send an email to yourself, and later on you can add it to your own collection of tips and points that you think you can learn and apply to your own life. These become part of your toolbox.

POINT OF VIEW

As I discussed earlier, success can be better achieved by having a positive attitude towards the things you set out to do as well as towards the things you don't want to do. As humans, we can often feel that we're getting into a rut or that life is getting stale, especially when the goals we're chasing take a long time to achieve. Even with a great desire to become really fit and lose weight, a 12-week fitness plan is still a hard activity to undertake and stay committed to, even when you're using self-discipline and keeping motivated. To avoid feeling like you're getting into a rut or like you're losing motivation and discipline, you need to consider changing your point of view. A point of view is about how you look at your tasks, activities, situations and endeavours. Your understanding of this is known as your perspective.

Your perspective on life generally or on something in particular will greatly affect the attitude you have towards it. Perspective can be influenced by previous experiences, other people's opinions, the position from which you're looking at something, or lack of awareness of the bigger picture. The sayings, *don't judge a book by its cover* and *you can't see the forest for the trees*, both refer to perspective. Your perspective will influence your attitude towards something. It can make something look very negative or very positive, whereas the opposite may actually be true. To get a realistic understanding of how something looks or whether your perspective is an accurate reflection of a situation, you need to try looking at it from multiple points of view.

Undertaking a 12-week fitness program appears very daunting if you simply see it as a single 12-week activity. The fitness program appears less achievable as you're focusing only on a single 12-week block of time and the huge number of physically intense gym sessions and the long period of restrictive eating. However, if you take a different point of view, perhaps by looking at only two weeks of the program at a time, the entire thing becomes more realistic and seems less daunting. You are able to visualise yourself undertaking and completing each session over a two-week block of time and see only the fortnight in front of you. Contrast this with a picture of never-ending self-torture that you will quickly lose motivation for. Start to look at it in the context of your whole life – 12 weeks is not that long a time, and several lots of two weeks will seem even less. You can focus on one milestone at a time. Your perspective will change, and your attitude will remain positive.

Additionally, varying the routine and location of your sessions helps change your point of view towards the training. Seeing the same equipment and the same faces day after day can become monotonous. You may need to literally change what you're looking at to keep motivated and disciplined. Go outside and do some body circuits instead of doing the workout in the same gym, or take a different route on your jog so that you see different landmarks. A simple change of scenery will help keep you on track. You will still be working towards your goal.

An infantry battalion utilises its reconnaissance and sniper teams to assist in developing the best possible information about the battlespace they're operating in. These highly trained soldiers are tasked to seek out information about an adversary or about the terrain that the commander intends to send forces into. However, a simple photograph or sketch image from one angle doesn't give the full picture, so the soldiers move to various positions to help get the complete picture. They seek to change their point of view so they don't, for example, miss the enemy tanks hiding behind the small hut they saw when they first approached a village of interest. Looking at something from multiple points of view gives them a more realistic understanding of the situation, and gives the commander a better perspective of the area they are operating in.

When the endeavours you're undertaking feel like they're becoming too difficult or insurmountable, look to change your point of view of the situation so that you can approach it in a different manner, from a different perspective and, in turn, with a renewed motivation. This will help you keep disciplined. Do this by looking at ways you can achieve the same thing but in a different manner. For example, consider working on a university assignment at the library if you usually work on it at home. You might find there are fewer distractions there. The simple change of scenery might make a difference in your approach to your assignment. Review the parts of a subject that you do know rather than dwelling on all the parts you don't know yet. Consider how much you knew when you started the course (likely nothing), and remind yourself of and appreciate the fact that you *are* actually learning the material. You simply need to keep repeating the same process (studying) so that the remainder of the course information is in your head. Don't take the 'glass half empty' perspective. The likely reality is that you're on track to success already. Remember to take a moment of pause to review how you're going, or you'll just burn yourself out.

Change your point of view on a large essay that appears daunting. Don't look at the entire assignment as 5,000 words. Look at it as several 1,000-word assignments so that you can view it tangibly and allocate realistic timeframes towards completing it. Don't allow one poor assignment result to influence your perspective. Sit back and look at the situation holistically and from other points of view. It may be that you completed that assignment while ill or trying to hold down a part-time job. Your poor result may simply have been an anomaly. It may have been the first time you were undertaking something of that type, size or scale, so learn the lessons and try again. The reality may be very different from your current perspective. If your perspective is accurate, however, then start looking at the steps you need to take to rectify the situation. It's far better to be going along in life with your eyes open, having proper situational awareness and accepting the realities of a situation than to be falsely telling yourself that everything is okay. You will not achieve success this way. Seek the assistance that you need, and maintain your resilience so you can keep progressing towards your goal.

If possible, consider mixing up your battle rhythm to help make it feel less like you're simply going through the motions and to prevent your routines from becoming stale. You can help change your perspective on your life in general by starting to do different things, or in a different order, which may help invigorate or remotivate you. Trying new activities will help you look at life differently, or at least allow you to feel like you're not stuck in your own little bubble, which may in turn reinvigorate your own motivations. Go and see a non-English-speaking movie. Treat yourself to a massage. Read a book of a genre you've not tried before. Try a cooking class, or get up early and go for a walk to see what the world looks like as the sun is coming up. Go and see an art exhibition, or try something cultured. A holiday, even if only an overnight stay in a hotel in your own city, can be refreshing and help change your perspective on things that may be troubling you. Anything that is out of the norm for you can expand your horizons and change how you're viewing things in your life.

Try listening to a different type of music or a new podcast on your journey home from work. This will help clear your mind and get you ready to deal with all the home issues by giving you the means to distance yourself from all the work issues. Perhaps consult colleagues or friends to get various opinions on a matter, especially if you think you've become emotionally entrenched in the situation. A clearer mind will help you make more informed decisions.

These foreign activities will naturally open your mind up, which will allow you to think about things more clearly as your mind is forced to take in new events and experiences. School teachers have been doing this for years, often taking their class outside on a sunny day to conduct a lesson. The mere change of scenery can help children think clearly and, more importantly, behave better. Likewise, the best soldiers are those who not only have physical capabilities but can also think critically and laterally and have an appreciation for many things. This is what makes them well-rounded individuals. They have the ability to think smart and act smart, too.

In this age of technological wonder, where many commercial interactions are made with credit cards or online, it can become easy to overspend on your budget because you're not actually using cash, which is more easily managed. Alter your point of view on your financial management by using only cash for a month (on discretionary spending at least, as most people have automatic payments for things like home loans). Looking at financial management from a different point of view will improve your perspective on it. Being able to actually see the cash go out and slowly whittle down over a month may help you to appreciate (or reappreciate) the value of a dollar and your use of it.

The workplace is an area that often has friction. An alternate point of view can allow you to resolve issues and deal with conflict in a more effective manner, or it can help you to view the situation differently to get a better understanding of whether your perspective is accurate or not. When dealing with a boss or a subordinate becomes difficult, remove yourself from the situation, take a breather and then resume dealing with the situation with a calmer and clearer mind. Due to the amount of time spent at work, which can be a third of a 24-hour day, you can become overly engrossed in the dynamics of the workplace, emotionally involved in work projects or drawn into workplace dramas. This can lead to a blinkered effect where it seems that work is the be-all and end-all of life and that the whinging from some employees is an accurate reflection of the competency of the boss, the colleague they are complaining about or the general morale in the office. These types of things lead to stress in the workplace, which leads to stress in your personal life. Thinking and acting smart is where you take a moment to remove yourself from the present situation and look more closely at the reality of where you are. You seek to look at it from multiple points of view so that you can get a better perspective. This will help you make fewer emotional and potentially irreversible decisions, such as quitting your job simply because you had a bad day.

At Red Diamond, when we consult with clients about looking at a situation from multiple points of view, we always advise them to ask themselves some key questions in a place where they feel comfortable and after they have removed themselves from the situation and had a break (e.g. at home at the end of the day). Is the boss really a terrible leader, or do some of your colleagues simply not like being told what to do? Are the demands of that work assignment coming from the boss, or have you put too much pressure on yourself unnecessarily? Are the battles you're picking really worth it (don't sweat the small stuff!), or are you not allowing yourself to see the other person's point of view, which may actually be the right one (or at least help reach the desired solution)?

Find a different point of view by looking at how you can do some things differently in order to remove those frictions or manage them better. If you're a leader, stamp out workplace gossip and negativity or speak to a trusted subordinate to get a better idea of what is actually going on. Try doing a gym session or going for a walk at lunch to break up your day (and to get away from other colleagues for a while). If you're a subordinate, have a conversation with your boss to clarify any concerns you have or to get a better understanding of where you stand in the scheme of things. You may have been misreading a perceived dissatisfaction in your performance all along. Make an effort to get to know a work colleague who may, in fact, just be shy as opposed to what everyone else perceives as rudeness.

The same theory applies to personal relationships. Because we devote big parts of our lives to our relationships, we become emotionally involved and attached. This is fine, but we still need to look at things from different points of view so that we can maintain a clear perspective and move towards the appropriate outcome.

There are plenty of stories and movies about people facing and overcoming difficulty and adversity which can help you view your own situation with more humility and perhaps inspire you to take a 'glass half full' approach. Despite any hardships we experienced on deployment and the difficult circumstances we faced, they often paled in comparison to the experiences and hardships faced by the civilians who lived in the areas we were operating in. The refugee camps and the makeshift hospitals, full of wounded and sick people, that we worked in and around were often enough to make even the most hardened soldiers appreciate their own lot in life much more significantly and made us all view things with a higher level of humility and with a different perspective. On any deployment the fact was, if all went in our favour, we would at some point end up going home and back to all the relative luxuries that being from a western nation had to offer. The perspective of most soldiers in relation to any perceived hardships they face is always tempered by the experiences they have had, and they know full well that someone else has it much tougher than they do, even if they themselves are someone else's 'have it harder' example.

MAKING A DECISION

No matter what we do in life, we all have to make decisions. They can be as small as what to eat for breakfast or as big as what type of mortgage to take out on a home. Decisions can be influenced by many things, the most predominant being the limitations (or options) available to you when making those decisions. If you have nothing left in the fridge and cupboard except for a few pieces of bread and some peanut butter, then making the decision about what to eat for breakfast is fairly easy. Similarly, with a home loan, your financial situation will dictate how much you can afford and what kind of mortgage you're able to take out. Making a decision is about finding a solution to a problem or an issue. (It can be a good or a bad problem or issue.) Making a decision requires that there be an option or a choice – one, few or many – in order to allow for something to happen and the problem or issue to be resolved.

Soldiers are required to make decisions on a daily basis, whether in peacetime or during war. The complexity of these decisions usually involves the rank level that each soldier has. For example, a four-star general makes daily decisions which affect the entire army. Junior soldiers, in peacetime, typically make daily decisions that affect only their section or platoon. In a combat or operational scenario, however, these decisions can have greater effects as they are made in environments where consequences may be lethal.

In the days prior to the D-Day invasion landings at Normandy beach on 06 June 1944, the Allied senior leadership, under Supreme Allied Commander General Eisenhower, were faced with deciding when to send a task force of hundreds of thousands of men, all waiting in various locations in the UK, across the English Channel and onto the beaches and fields of France. The Germans knew an invasion was imminent – although the location was successfully kept secret up until the landings – and the sheer size of the operations meant that failure was a distinct possibility. Eisenhower even had a statement ready to deliver to the press if the landings, codenamed Operation Overlord, failed. The weather, always a significant factor in military operations, was slowly deteriorating and had already caused a 24-hour delay (the initial planned date was to be 05 June), and Eisenhower was cautioned by his subordinates that continued delay would undermine that chances of the landings' success, as plans had taken into consideration factors like tide levels and moon phases, all of which were about to change. Eisenhower, the sole decision maker as to when the invasion would occur, had to quickly determine when the landings would take place. The risk was that the invasion would be hindered by Mother Nature and had the potential for tens of thousands of allied casualties.

During the war in Afghanistan, the ACMs often hid amongst the local population as a means of disguising their activities, usually by pretending to be farmers. They often waited, hidden from view, until coalition patrols approached the village before picking up their weapons and firing at the approaching soldiers. Then they dropped the weapons and picked up the farming tools again. Standard procedures followed by both US and Australian soldiers when fired upon, known as the Immediate Action (IA) drill, require that soldiers find protection (cover) from the incoming bullets and determine what is happening. The soldiers have to quickly determine whether to return fire, and the patrol commander has to decide whether to keep moving towards the village and engage the ACM at close quarters, possibly while in close proximity to civilians. The risk was that civilians could be killed in the ensuring firefight, and the ACM would use those deaths as propaganda to turn the local support against the coalition. The decision ultimately made would be based on the requirements of the mission and weighing up the risks involved.

In both those scenarios, a decision had to be made. While both consequences were potentially lethal, the numbers of soldiers involved and the potential effects were diametrically different. However, the basic principles and processes used to make a decision by both Eisenhower and the patrol commanders are essentially the same. The army uses various decision-making processes, which are easily applied to any scenario, to help make decisions ranging from what training activities soldiers will conduct during any given week right up to how to invade a country.

The army uses the Military Appreciation Process (MAP) to assist in planning and decision making. There are several levels of MAP, each increasing in complexity based on the level of planning required, the time available and the number of planners involved. However, in all of them, there are four key steps:

1. Mission Analysis
2. Course of Action Development
3. Course of Action Analysis
4. Decision and Execution.

No matter how big or how small the situation, this tool enables better decision making as it follows a process which allows the decision maker to consider the factors and options which need to be taken into consideration before making a decision. It allows for a more considered decision to be made and helps a commander remove the ambiguity that often comes with military operations. I won't endeavour to explain the entire MAP in detail – the current doctrine alone reads like a novel and is a highly complex instrument – so therefore I'll provide the key points.

1. Mission Analysis:

The decision maker will first determine if a decision has to be made at all. If the army is doing the MAP, then the answer is yes, because the army only undertakes the MAP in response to being given a mission to plan. In mission analysis, the current situation is looked at closely to determine what needs to be done (i.e. what needs to be achieved). This helps clear up ambiguity and confusion. I'll use the scenario of buying a car as an example. The situation is that you need a car as the buses don't go anywhere near your new workplace, so public transport is no longer an option. The first action is to determine how much time is available to make the decision. From here, a soldier or a commander seeks as much intelligence as possible to better understand the situation. Some situations call for an immediate decision, whereas others may allow for months of planning and research to aid the decision making. If you start the new job within a week, you need to buy a car soon. If you're going to start using the second family car as the work car and plan to buy a nicer car as the family car, then you have more time to shop around and find bargains. From here you can start looking at the options available to help you to make a decision.

Every infantry battalion has a reconnaissance (recon) platoon whose sole job is to be the 'eyes and ears' of the commander. These soldiers move out onto the battlefield first and seek to gather intelligence that the commander needs to answer questions about enemy dispositions, terrain composition and anything else that may help with the planning so that a more informed decision can be made. The wider military has many assets used for recon, ranging from submarines to highly digitised satellites that can take high quality images from space. The purpose of recon is to increase the amount of intelligence commanders have available so that they can make a more informed decision. When starting out with the MAP, recon will be one of the first activities conducted.

When required to make a decision, be it a big one or a small one, you need to conduct your own recon so you can be better informed to make that decision. Your recon may range from something as simple as opening the local newspaper to see what cars are for sale right up to getting multiple quotes for a loan, comparing prices on various websites and conducting research on the various pros and cons of the available options.

2. Course of Action (CoA) Development:

In this step, the decision maker will examine some of the possible decisions that *could* be made. Having already considered the mission and having gathered as much information as is available, the decision maker now looks at what options are available. Soldiers use the intelligence available and start to consider what they can do within their resource limitations. They look at whether those options are actually feasible and suitable to the requirements of the mission and, if not, they will exclude them from consideration. It's not feasible or suitable to buy a shiny red sports car if it's out of your budget and doesn't fit the whole family for a Sunday picnic. You may have only a few options available based on your financial resources and the requirements of your family. The army always tries to come up with three options to help make a decision, especially if there is a decent amount of time available to research those options. This allows debate amongst staff and helps find the *best* option as opposed to the *okay* option. However, the situation may present only one option. Patrol commanders may be bound by ROE where they cannot return fire if there is any doubt about civilians being in a village (as happens frequently), so the decision is made for them. If the local car yard has only one seven-seater car in your price range, then the decision may be already made for you.

3. Course of Action (CoA) Analysis:

In this step, the decision maker must look at the possible options that were determined in the last step, consider and analyse their benefits and consequences and look at whether those options will lead to an achievement of the mission – if one of them will achieve the desired outcome. It's a closer look at the options to get a better understanding of how making a particular decision will end up working out. Eisenhower, with all the ships, planes and personnel under his command, had only a few options from which he had to make a decision with regards to conducting the landings in different locations and at different times. With the support of his staff, he examined closely what those options were to get an appreciation of what the consequences might be of each option and whether it would allow the mission to be a success. Similarly, you have to look closely at the new car options available to you to get a better idea of which one achieves your aims of getting you to work *and* driving the family around. Further consideration may show that a second-hand car may be cheaper now, but the ongoing running costs make it a more expensive option in the long run, so that is a negative consequence of that option. You must look at the pros and cons of each option.

4. Decision and Execution:

The final step of the MAP is actually making the decision and then doing whatever is needed to undertake the requirements that the particular decision calls for in order to achieve the outcome. Eisenhower made a decision and decided to conduct the landings at Normandy on 06 June 1944. There was still significant risk that went with that decision, but it was made based on the intelligence available, with consideration given to all the pros and cons, and it achieved the mission which was to land allied forces on the European continent to open up a Western Front to fight the Germans. Similarly, once you have considered and analysed your car options, you need to head to the car yard and hand over the cash.

The MAP can be used for any decision-making requirements. The bigger decisions in life often require more time and a detailed consideration of the information and options available before a choice is made, such as whether to accept a new job, buy a house or move interstate for a new relationship. You must always keep in mind one key question: *what outcome are you trying to achieve? (i.e. the reason you need to make the decision)*. Smaller decisions may require less thinking and usually come to down to common sense and are based on previous experiences. For example, going to see the late session at the movies or out drinking with your friends on a weeknight may not be a good idea if you need to be up at the crack of dawn for a job interview, especially if you know that you're not a morning person.

Soldiers under fire have a very limited timeframe in which to make a decision, and the consequences of making the wrong one can have very damaging results. Even so, they will use the limited time available to conduct a quick appreciation of the prevailing situation and quickly think through the MAP to help make an informed decision. The MAP has a benefit in that it forces the mind to focus on the process as opposed to the emotion of the situation. We're all only human, so decisions will always be influenced by feeling, gut instinct and personal preference. The MAP is simply a way of making a more informed and considered decision and reducing indecisiveness. It always pays to run decisions by friends or colleagues, as they will be able to give opinions or considerations you may not have thought about when going through the process by yourself. However, you must always remember that, when you make a decision, you need to be prepared to be accountable for it, so it pays to try and make the right one the first time around.

Another way of looking at the MAP is the *4-step process*. This is virtually identical to the MAP but is laid out using more common terminology:

1. **Issue** (Mission Analysis) – Why does a decision need to be made and what is the desired outcome?
2. **Options available** (CoA Development) – What options are available based on your capabilities and resources?
3. **Pros and Cons of each option** (CoA Analysis) – Which option will achieve the outcome and be the best for the situation and your needs?
4. **Decide** (Decision and Execution) – Be decisive and make the decision.

Soldiers use a basic tool, called a Quick Decision Exercise (QDE), to practise assessing a situation and making a decision. QDEs are taught to officer cadets and junior soldiers as a means to become proficient in military thinking as all QDEs are based on a military scenario, either real or imagined. A scenario is presented to a group or an individual – usually a military situation that is based on the soldier's trade or capabilities – and they are then given a small amount of time in which to make a decision and outline their course of action. They are asked to justify their decision and give reasons for it, so they use the MAP as a means of helping them to arrive at the decision. Some scenarios are grey in nature and are designed to encourage and develop thinking and theory skills, whereas others have clear-cut right and wrong answers. You can incorporate QDEs in your workplace or other areas, such as a sporting team, to help develop your ability to think quickly and make decisions under pressure. Using the MAP and similar processes, as well as conducting QDEs, will increase your own confidence to make decisions and remove indecisiveness from the decision-making process.

THINK AHEAD

Planning is one of the key skills taught to all soldiers, regardless of their rank level or chosen trade. Planning allows the army to anticipate and prepare for any task, ranging from conducting PT at platoon level up to high-intensity combat operations. Planning forces soldiers to think about a situation objectively, consider what may happen and what they may need to do when preparing to conduct that task. One of the benefits of undertaking the MAP as part of the decision-making process is that it allows you to think about the outcomes of the available options so that you can make an informed decision that considers the result of your subsequent actions. To do this well, you always need to think ahead and have situational awareness.

Soldiers have a term called *switched on*. Being switched on is a term of respect and can be used individually and collectively. It refers to the high level of discipline, professionalism and proficiency within an individual or group of soldiers. To be referred to as being switched on means that other soldiers have identified that an individual or group has achieved, and maintains, a high level of discipline, professionalism and proficiency in their job. It is one of the highest informal accolades soldiers can bestow on each other. The opposite term, one used particularly in the Australian vernacular, is to have your *head up your arse*. This means that an individual or group has poor discipline, lacks professionalism and is incompetent at their job. They lack an understanding or appreciation of what is occurring around them and are generally regarded as hindrances to efficient military operations.

These two terms are not synonymous with the army, however. We've all seen people, either in the workplace, at the gym or in the community at large who conduct themselves with complete disregard to the people or environment around them. They obviously have made no effort to think ahead or consider the people around them. They're the type of people who walk the streets with their heads down, looking at their phones and without looking around them, and then complain when they bump into someone.

Think about when you get onto a plane to fly somewhere. Airports are inherently frustrating places, as there are many people all trying to check in, get through security, board their plane and take their seat for take-off. There are many friction points in the process, but they are always exacerbated by the people who have their head up their arse. These are the people who get to the check-in counter, with a massive queue behind them, without their identification or passport ready to present to the counter assistant, and have to dig into the bottom of their suitcase as opposed to their wallet to get it. They're the people who walk through the security point, look annoyed when the metal detector goes off and then proceed to remove a small silver mine's worth of metal from their clothing. They're the people who get to the ticket point, same queue behind them, and haven't got their ticket ready to present to the flight attendant even though every single person in front of them has managed to complete the process. These are the same people who decide that, as the other passengers are trying to find their seats, they want to get up and block the aisle so that they can dig into their bag in the overhead compartment to retrieve a book for the flight, which they don't read anyway because they decide they want to talk loudly during the flight even though all the lights are out and the rest of the passengers are trying to get some sleep. The odd first-time traveller notwithstanding, who isn't familiar with the proper routines at airports, these types of people are definitely not switched on. They lack any form of self-awareness which means they haven't bothered to try and understand the situation or the environment they are in or done any level of basic planning.

The opposite types, those who are switched on, are prepared when they get to the ticket counter. They have all their belongings ready to place on the tray to pass through security; they are ready to present their ticket to the flight attendant at the boarding gate; and they have any items they want ready for when they sit down, or at least they wait until the majority of the people sit down and, at worst, only briefly disrupt the person ahead of them.

This phenomenon can apply to many scenarios. Naturally rude and obnoxious people aside, the delays are caused because people have not thought ahead and planned out what they need to do for that situation.

Even at the lowest tactical level, soldiers always think ahead. As infantrymen, we were taught to always plan for the next fire position. When in contact (a battle) with the enemy, soldiers look for cover to seek protection from the incoming fire and to allow them to assess the situation and await orders from the commanders. The decision may be made to make an assault to repel the attack; they may be the ones initiating the attack on an enemy position. A fire position is a place from where soldiers are firing their weapons during an attack, somewhere that affords them protection from incoming fire but also allows them to accurately fire their weapons. During an infantry attack, soldiers are constantly moving forward so that they can get closer to engage the enemy and capture their soldiers or assets, such as a gun pit or a building. At all times, they scan the ground in front of them and look for the best place to take their next fire position, always maintaining an awareness of their fellow soldiers around them so as to avoid any incidents of fratricide. They are always thinking ahead, if only in anticipation of the next few minutes of action. The commanders in the battle are thinking about what they need to do once they win the firefight, as they need to consider things like counter-attacks or the need to get medical support for wounded soldiers.

Thinking ahead, which is essentially just forward planning, allows you to be more successful because you will be in a good position to exploit the opportunities that present themselves in your life and because you will not waste time dealing with situations that arise as a result of poor organisation. Thinking ahead allows you to be prepared for the eventualities that can occur unexpectedly. This holds true for the small things, like having packed a warm jacket to a sports game that is held outdoors at night, to the larger things, such as having a career progression outline in place so you can complete all the pre-requisites required to receive the next promotion.

Chess is a strategy game developed in India during the sixth or seventh century which has been used for centuries to teach military professionals the art and science of forward thinking. Skilled chess players always have to consider that they have an objective (to win the game) and that each move they make with their pieces has a consequence. They must consider that the move they make now may have consequences (both positive and negative) several moves from now. Therefore, they develop a plan of attack and are prepared to amend that plan as needed when the situation changes. The best modern-day example of an excellent tool to develop this skill is the video game Tetris.

In your own life and endeavours, you need to always think ahead as to what may occur as a result of your decisions or what you need to do in order to be prepared to undertake something you want to do. For example, a course of university study typically requires that several courses of a certain subject need to be completed as a pre-requisite for going on to undertake a higher level of learning. You need to consider this when choosing your courses. Something as enjoyable as going on holiday means you have to ensure you have obtained a visa to enter the country of your holiday destination. It pays to do some research as to what is involved to ensure that you don't leave it until the last minute when the travel office may be closed. Even the simple things require thinking ahead. Over time this becomes second nature anyway, but at first you need to give some consideration to them. If you're going to head to the gym straight from work, you need to ensure you have all the equipment appropriate to the session you're intending to do so that you don't have to waste extra petrol (and time) going home to get them first.

Thinking ahead will soon become a natural part of the way you do things. In everything you do and in the decisions you make, you will not only be considering the present but also the near and long-term future. Doing so will ensure you don't waste unnecessary time and effort and will allow you to ultimately be more successful in your endeavours.

SOLDIERING IS A LIFESTYLE

Perhaps the best method that soldiers use to achieve success is the fact that they don't consider soldiering just a job, but they treat it as a lifestyle. Like professional athletes, the best soldiers completely devote themselves to their trade, making the necessary sacrifices to achieve the high levels of proficiency needed to be successful in combat. Football teams seek to win premierships and titles; investment bankers seek to make money; doctors seek to be able to cure diseases. All these result-driven ambitions are achieved through a complete commitment to the task, and all its associated requirements, in order to achieve success. These types of trades are considered professions as opposed to jobs because they inherently require long hours of training, study and practice in order to become proficient at the skills required of them. Soldiering is the same.

Even when on leave from duty, soldiers still go to the gym to keep in physical shape, and they still study to prepare for the promotion and trade-specific courses they are aiming for. They still study intelligence reports to be aware of emerging trends in weapons advances, as well as current affairs, so that they are fully aware of the things they may be expected to face while on operational deployment. They still make time for family and a social life, but they make greater sacrifices in how much time they devote to them – far greater than what is expected of most civilians. Soldiers cannot just turn their phones off when they finish normal duties for the day and expect that they can just do what they want. Normal sacrifices required of soldiers include restrictions on how social media is used and the type of people they can associate with, especially if they hold a security clearance. At any given time, an Australian infantry battalion is 'online'. This means that they are prepared to deploy at very short notice to any location that the government deems they are required. This means they sacrifice the ability to head away for a weekend with the family or the amount of alcohol they can drink on a Saturday night. They live their lives with the main focus being their profession. My first ever military deployment involved arriving for what I anticipated to be a normal training day, being deployed at short notice that afternoon and then getting back home four months later.

Soldiers treat their profession as a lifestyle so that, when they are called into battle, they go with peace of mind knowing that they have done everything they can to be as conditioned and ready as possible to undertake the tasks required of them. As I stated earlier, soldiering can be unforgiving, and it doesn't accept mediocrity so treating it like a 9-to-5 job is not tolerated.

Achieving success in your own life is about determining if the lifestyle you're currently living is conducive to achieving the goals you have set for yourself. Before setting goals, you must ask yourself what you need to do and what sacrifices you're willing to make in order to achieve them. Combat soldiers, life professional athletes, completely devote themselves to their trade. It not only becomes their lifestyle; it becomes their life. You don't need to make work your entire life if you also want to focus on family, fitness or a social life. You simply need to establish a lifestyle that allows you to achieve those goals. There is no point undermining your ambitions by not doing the things required and by not making the necessary sacrifices that will allow you to achieve them. Losing weight, for example, requires a change to better nutritional habits as well as time in the gym, so you need to eat out significantly less and start eating healthy. Less time on the couch and more time in the gym is also a lifestyle requirement needed to achieve this goal.

When I was studying to become a high school teacher, I was also working full-time at a service station. My lifestyle had to be one that allowed me to do these things. As much as I would have loved to stay out late on Saturday nights with my friends, I made sure I was home at a decent hour so that I could be up to work the next day and in good condition to study in the afternoon.

Setting a battle rhythm (a personal routine) dictates when you do each of your activities, whereas the type of lifestyle you lead determines the type of activities you will do. Your lifestyle will alter and adapt to your changing ambitions. When you're younger, your lifestyle may be one where you can stay out late most nights and be able to get up at a decent time the next morning. As you get older and perhaps marry and have children, your lifestyle will change to focus on your family as opposed to going out all the time. Similarly, small lifestyle changes may need to be altered and adapted to facilitate new goals and ambitions. Non-professional athletes who want to run a marathon must set out to specifically train for such a demanding event. Therefore, they become more focused on running training, which requires a sacrifice of weekend sleep-ins. Such sacrifices are required right up to the day of the big run. A lack of commitment and dedication to that goal will result in not being able to run the entire length of the course.

Setting a lifestyle is about doing the things needed to allow you to achieve your goals and committing yourself appropriately. Far too many people these days, it seems, expect instant results without doing the associated hard work. Your personal endeavours don't have to be the entire focus of your life unless you want them to be (as soldiers and professional athletes do). Lifestyle is about giving yourself the best possible opportunity, within your own specific life circumstances, to set out and achieve success in your personal endeavours.

SUMMARY:

- **Soldiers think and act smart as a matter of routine. They implement basic concepts to allow them to do this and to achieve success:**
  - **Failure is not an end state – have a Plan B**
  - **Never stop learning**
  - **Making a decision**
  - **Think Ahead**
  - **Soldiering is a lifestyle**

Conclusion

"The soldier is the Army. The soldier is also a citizen."

— General George S. Patton

The skills and knowledge that soldiers acquire over their careers and that they apply to their subsequent civilian life are not rocket science. They are simple yet effective concepts that have been proven in combat and in peacetime. Many soldiers have swapped their camouflage uniforms for the suit and tie of the boardroom, the high-visibility clothing of the building site or the labelled polo shirts of small business. They have taken their skills and successfully applied them to new environments, remaining motivated, diligent and mission focused. It's these skills that are highly sought after and that are now being seen among veterans as they walk the halls of power in Washington and Canberra or lead companies into exciting futures creating new technologies. Their application to your own life will enhance your prospects of personal and professional success as they have been proven to work.

However, any successes or failures that occur in your life still come down to one key, irrefutable factor: YOU and your own willingness to develop professionalism, to instil discipline, to enhance resilience, to recover from failure and give something another go. These concepts allow you to be better prepared to deal with the challenges in your life and to confidently undertake the tasks that will allow you to achieve your goals. You don't have to be a soldier to utilise the skills that the army uses to achieve success. You can leverage the lessons they have learnt and apply them to the areas of your own life that you want to develop and improve on. It may simply be about approaching your gym routine in a more professional manner or making decisions in your workplace with more consideration to what the potential consequences might be. Take and apply these lessons in the way that suits you best.

People will quickly begin to notice when you start thinking and acting in a more professional manner and will seek to know the secrets to your methods. It's not complex. These concepts are easily digested and applied to anything that you want to do, either holistically in your life or to specific pursuits. Your measure of success is still defined by you and the things you want to achieve. Give yourself every opportunity, be honest with yourself and don't ever quit!

Please feel free to follow us on social media and
provide recommendations and feedback!

INSTAGRAM

joshfrancis_red.diamond

FACEBOOK

joshfrancisbooks

AMAZON

search for **The Camouflage Series**

Please leave an honest review on Amazon. This helps to tailor better
content and allows for reader interaction.

Sign up to the readers group at www.red-diamond.com.au/books

Biography

Josh Francis qualified as high school teacher before commissioning into the Royal Australian Navy as a junior officer soon after the September 11 attacks in the U.S. A desire to serve on warlike operations saw him resign his commission and enlist into the Australian Army. After qualifying as an infantryman and paratrooper, Josh deployed on peacekeeping operations in Timor-Leste conducting counter-militia operations.

After completing basic and specialist intelligence operations training, Josh completed multiple deployments to Afghanistan and Iraq, conducting duties in conventional and special operations, as well as training roles.

He is the author of the military themed personal development books *The Camouflage Series*, as well as the *Zach Kryton* series of books.

www.ingramcontent.com/pod-product-compliance
Lightning Source LLC
Chambersburg PA
CBHW050510210326
41521CB00011B/2405